Antonio Eudes de Sousa Oliveira
Francisco E. P. Mousinho
Regina Lúcia F. Gomes

Effect of water deficit on fava bean development

Antonio Eudes de Sousa Oliveira
Francisco E. P. Mousinho
Regina Lúcia F. Gomes

Effect of water deficit on fava bean development

Water-soil-plant-atmosphere relationship in a protected environment

Imprint

Any brand names and product names mentioned in this book are subject to trademark, brand or patent protection and are trademarks or registered trademarks of their respective holders. The use of brand names, product names, common names, trade names, product descriptions etc. even without a particular marking in this work is in no way to be construed to mean that such names may be regarded as unrestricted in respect of trademark and brand protection legislation and could thus be used by anyone.

Cover image: www.ingimage.com

This book is a translation from the original published under ISBN 978-613-9-63895-6.

Publisher:
Sciencia Scripts
is a trademark of
Dodo Books Indian Ocean Ltd. and OmniScriptum S.R.L publishing group

120 High Road, East Finchley, London, N2 9ED, United Kingdom
Str. Armeneasca 28/1, office 1, Chisinau MD-2012, Republic of Moldova, Europe
Printed at: see last page
ISBN: 978-620-7-71609-8

"I do not know how I look in the eyes of the world, but even I see myself as a poor boy who was playing on the beach and had fun finding a smoother pebble once in a while, or a prettier shell than usual, while the great ocean of truth lay all unexplored before me. "

Isac Newton

To my father Joaquim de Oliveira and my mother Maria do Socorro To my grandfather Torquato and my grandmother Maria Gomes To my siblings Marcos, Mazim, Jose and Iamara

DEDICATE

To my wife Patrícia de Sousa To my son Vinicius de Oliveira

OFFER

ACKNOWLEDGEMENTS

To God, the provider of all good, for my existence and the strength to achieve my goals, because without him nothing is possible.

To the Federal University of Piauí, for giving me the opportunity to study for a master's degree in Agronomy and to carry out all my work.

To the Agricultural College of Teresina, for their help and support in setting up and conducting the experiment.

To Prof Dr Francisco Edinaldo Pinto Mousinho, for his support, his efficient guidance, his understanding, his knowledge, his friendship, for all the teaching and knowledge he has given me since I was an undergraduate and for all the patience and trust he has placed in me throughout this time.

To Professor Dr Regina Lúcia Ferreira Gomes, for her friendship, guidance, patience and understanding.

To the Director of the Prof José Bento Agricultural College, for his support in carrying out the experiment and for his friendship.

To the other professors of the Master's Degree in Agronomy at the PPGA, for their teachings, which were of fundamental importance for the preparation of this work.

To Vicente de Sousa Paulo, Secretary of the Master's programme in Agronomy, for his respect, conversations and joys.

To my dear wife Patrícia de Sousa, for helping to smooth out this long journey, for her love, dedication and understanding during times of stress and absence;

To my fellow master's students, Gisele Castelo Branco, Dorotéia Marçal, Maiany Carvalho, Antonia Maria, Ricardo Silva, Sergio Augusto, José de Ribamar, Theofilo Santos, Chico Léo, Sabrina, for all the fun and fellowship in and out of the classroom;

To my long-time friends, Agenor Rocha, Natalia, Wilon, Kadson, Jacquecilene Moura, Marcelo Simeão, Minguel, Michel Barros, Dyego Ferreira, Diego da Paz, Jedeias Amorim, Bruno Guerra, Marlon Castelo Branco for their friendship and fraternal coexistence at all times, especially when we were relaxing at "Mazé".

To my parents Joaquim de Oliveira and Maria do Socorro, and my grandparents Torquato and Maria Gomes, for their example of struggle, determination, fibre and serenity, and especially for having been my reference point in the quest to be a citizen, for the empirical knowledge passed on and the motivation to seek scientific technical knowledge. To CAPES for the research grant.

To those who, directly or indirectly, contributed in some way to this work.

My eternal thanks.

SUMMARY

SUMMARY

The aim of this study was to assess the effects of water deficit on the development of fava beans, as well as to determine the coefficient of sensitivity to water deficit (Ky) in the phenological phases in a protected environment. The experiment was conducted in pots under a greenhouse with a 15 *pm* thick low-density polythene cover in the experimental area of the Teresina Agricultural College of the Federal University of Piauí (UFPI), Teresina-PI. The experimental design was entirely randomised (DIC) and the treatments consisted of a combination of inducing water stress of 50% of the fava bean's evapotranspiration: water stress in vegetative phase I; water stress in vegetative phase II; water stress in reproductive phase III; water stress in reproductive phase IV; water stress in vegetative phases I and II; water stress in vegetative and reproductive phases II, III and IV; Water stress in reproductive phases III and IV; Water stress in vegetative and reproductive phases I, II and III; Water stress in vegetative and reproductive phases I, II, III and IV and a treatment without water stress. The number of pods per plant, number of grains, weight of 100 grains, grain and pod yield, leaf area index, total dry matter as well as the sensitivity factor to water deficit and water use efficiency were evaluated. Water deficit significantly affected leaf area index, chlorophyll index, number of pods per plant and increased flower and pod abortion. The fava bean was more sensitive to water stress when it occurred at more than one stage of development (over a longer period of time) and when it occurred during flowering and pod formation; it was more tolerant during vegetative development and pod ripening. Water deficit affected fava bean yields in terms of pods, grains and total dry matter. The efficiency of water use by fava beans was higher in treatments without water stress.

Keywords: *Phaseolus lunatus* L., Water deficit sensitivity factor, Water use efficiency

CHAPTER 1

INTRODUCTION

Agriculture is an activity that presents a relatively high degree of risk, due to the inherent characteristics of the crops, the soil, the climate, the economy and the limited use of technology. Especially in the semi-arid zone of north-eastern Brazil, the occurrence of prolonged periods of drought resulting from the scarcity and poor distribution of rainfall, combined with socio-economic and cultural factors that partly limit the use of certain technologies by producers, transforms rainfed agriculture into a high-risk activity with low economic returns. Even so, when rainfall is sufficient and regularly distributed, agricultural production reaches reasonable levels. These conditions show that water is a limiting factor and that making better use of it, as well as using crops that are better adapted to these conditions, would result in an increase in crop yields. Therefore, water availability is an important factor when it comes to maximising agricultural productivity gains.

The irrigation technique, if used properly, could help to guarantee satisfactory crop yields. The use of irrigation, as well as determining the amount and timing of water application, is part of a decision to be made based on knowledge of water-soil-plant-atmosphere relations. To this end, it is necessary to study irrigation management methods and techniques that make adequate minimum amounts of water compatible with the requirements of the stages of development (phenological phases), as well as the influence of water deficit on productivity at these stages.

It is necessary to know the behaviour of each crop as a function of the different amounts of water supplied to it, the stages of its development that consume the most water and the critical periods when a lack or excess would result in a drop in production (BERNARDO, 2005). The water stress that develops in any particular situation in the plant is the result of a complete combination of soil, plant and atmospheric factors, which interact to control the rate of absorption and loss of water (VAADIA et al., 1961), although according to Gavande (1976), the response of plants to the "water" factor seems to be more closely related to the total water potential in the soil.

The fava bean stands out as one of the crops grown in the Northeast region of Brazil, cultivated under rainfed conditions, with little use of technology, by family farmers, resulting in low productivity rates and great oscillation in production. The sensitivity of fava beans to water deficit in the soil and climatic uncertainties, especially those related to rainfall variations between years and cultivation sites, determine the low yields and fluctuations in annual production of this crop.

According to Doorenbos and Kassam (1994), the relationship between crop yield and water supply can be determined when it is possible to quantify, on the one hand, the crop's water requirements and the effects of water deficits and, on the other, the crop's maximum and actual yield. These authors also emphasise that water stress (represented by relative evapotranspiration deficiency) and relative production are related by means of a coefficient (Ky), called the "response factor". This coefficient varies according to the different stages of crop development (vegetative establishment, flowering, pod formation and ripening). The "response factor" (Ky) is useful in the planning and operation of irrigated areas, as it makes it possible to quantify irrigation water and its use, in terms of yield and total crop production, for the entire irrigated area.

Considering the need for more information on the behaviour of the fava bean crop in relation to the application of irrigation, and also given the importance of the effects of water management on the production of the fava bean crop, the aim of this study was to evaluate the effects of water deficit on the productive behaviour of the fava bean, as well as to determine the coefficient of sensitivity to water deficit, the "response factor" (Ky) in the phenological phases in a protected environment.

CHAPTER 2

LITERATURE REVIEW

2.1.The fava bean - general considerations

The genus Phaseolus originated in the Americas and is widely distributed throughout the world, being cultivated in the tropics, subtropics and temperate zones (GAITAN-SÓLIS et al., 2002). The genus Phaseolus has around 70 species (FREYTAG and DEBOUCK, 2002), of which only five are cultivated by humans: P. vulgaris L., P. lunatus L., P. coccineus L., P. acutifolius A. Gray and P. polyanthus Greenman. Due to its economic importance, this genus has been the subject of various agronomic, systematic and molecular studies (WETZEL et al., 2006). According to Ramalho et al. (1993), P. vulgaris L. and P. lunatus are the most important species in the genus.

According to Melchior (1964), the genus Phaseolus belongs to the order Rosales, subtribe Phaseolinae, tribe Phaseoleae, subfamily Papilionoideae and family Leguminosae, however, Cronquist (1988) classifies it in the subclass Rosidae, order Fabales and family Fabaceae. Broughton et al. (2003) state that this family is one of the largest among the dicotyledons, with 643 genera and 18,000 species distributed throughout the world, especially in tropical and subtropical regions.

According to Silva et al. (2002), studies indicate that the number of species can vary from 31 to 52, all of which originate from the American continent. As for classification based on floral morphology, Delgado Salinas (1985), quoted by Debouck (1999), suggested four sections for the genus: Chiapasana, Minkelersia, Xanthotricha and Phaseolus. These groupings, according to Silva et al. (2003), were later confirmed through studies based on chloroplastidial DNA polymorphism and DNA sequences.

The fava bean (Phaseolus lunatus L.), one of the five cultivated species of the Phaseolus genus, is a tropical legume characterised by high genetic diversity and high production potential, which adapts to the most diverse environmental conditions, but develops best in the humid and hot tropics (MAQUET et al., 1999), being widely distributed throughout the Americas (GUTÍERREZ-SALGADO et al., 1995). According to Vieira (1992), the fava bean adapts best to fertile, well-drained sandy-clay soil, with good yields at a pH between 5.6 and 6.8, but it tolerates the most diverse environmental conditions and is considered to be more drought-tolerant than the common bean.

It is a multi-annual species, predominantly autogamous, with approximately a 10 per cent natural crossing rate (HARDY et al., 1997). According to Beyra and Artiles (2004), the growth habit of this species can be indeterminate climbing, with the terminal bud developing on a guide, or determinate dwarf with complete development of the terminal bud on an inflorescence. According to Zimmermann and Teixeira (1996), P. lunatus can be identified as an epigean germinating legume;

with trifoliate leaves generally showing a dark colour, more persistent than in other species of the genus, even after the pods have ripened; small, pointed bracteoles; pods that are generally oblong and recurved, with two distinct heights (ventral and dorsal) and the number of seeds per pod varying from two to four. These seeds vary greatly in size and tegument colour (SANTOS et al., 2002).

The fava bean is an alternative source of income and food for the population of the north-eastern region of Brazil, who consume it in the form of dried or green beans, and it is cultivated mainly by family farmers, who mainly use cultivars of indeterminate growth (OLIVEIRA et al., 2004). According to IBGE (2010), 7,349 tonnes of dried fava beans were produced in Brazil, in a planted area of 29,825 hectares. The states of Paraíba, Rio Grande do Norte, Ceará, Pernambuco, Piauí, Sergipe, Maranhão and Alagoas, in descending order, are the largest producers, and together they make the Northeast the largest producing region with production of 6,667 tonnes on 28,628 hectares. Piauí produced 485 tonnes of fava beans on 2,107 hectares in the 2010 agricultural year.

Fava bean cultivation has received little attention from research and extension organisations, resulting in limited knowledge of its agronomic characteristics. In several municipalities in north-eastern Brazil, fava beans are of relative economic and social importance, being a source of alternative food along with maize, cassava and cowpeas. Considered to be more tolerant of drought, excess humidity and heat than the common bean (VIEIRA, 2002), its cultivation in this region is rustic, in consortium with maize, cassava or castor beans, taking the plants of this crop as support (AZEVEDO et al., 2003).

According to IBGE (2010), a historical series of fava bean production data from the 1990 to 2010 agricultural year, productivity is low, ranging from 200 to 476 kg ha^{-1} , fluctuating with instability in production and productivity. There are also fluctuations in the area planted and the area harvested, with a 26.51 per cent decrease in the area used to grow fava beans in the north-east of Brazil between 1990 and 2010. In Piauí, the area under fava bean cultivation has remained constant over the last two decades, and low productivity rates have been recurrent. The low level of technology employed also contributes, as there is a lack of recommendations for adapted fava bean cultivars with production stability.

Phenology is the study of the periodicity of climatic conditions influenced by soil and ecological conditions in general, on the biological cycle of plants, especially the reproductive organs and vegetative growth (WIELGOLASKI, 1974). Knowledge and understanding of phonological patterns are essential for management and crop care. Factors such as water stress, excessive phosphate fertilisation, high temperatures, low light levels, inadequate photoperiod and biotic factors contribute to variation in the length of the reproductive period and low efficiency in the fruit/flower ratio (SUMMERFIELD et al., 1985).

Oliveira et al. (2010), studying the development of fava beans, observed that germination occurred between the sixth and tenth day after sowing, the cotyledons emerged between the ninth and fifteenth day, the first unifoliate leaves appeared between the thirteenth and fourteenth day, the first trifoliate leaf emerged between the sixteenth and eighteenth day, the second between the twenty-first and twenty-fourth day, the primordia of the secondary branch emerged between the twenty-seventh and twenty-eighth day. In the reproductive phase of the fava bean, the first flower buds appeared between 36 and 52 days after emergence, the first flower between 46 and 60 days, and the fruit ripening phase begins between 51 and 74 days, with 50 per cent maturity of the pods between 67 and 85 days.

Câmara (1997) reports that the length of time between the different stages of plant development can vary depending on the cultivar, temperature, climate and sowing time, among other factors.

2.2   Water-soil-plant-atmosphere relations

The development of plants depends on water to maintain turgidity and to cool the leaves. When the water supply is insufficient, the stomata close. Water stress reduces the elongation of plant cells, while stomatal closure reduces the availability of CO_2 and thus the production of assimilates and growth (KUIPER, 1961; JARVIS; DAVIES, 1998). Water moves through the soil-plant-atmosphere system by mass flow, mainly in the liquid phase (in very dry conditions, by vapour flow in the soil, where it plays an important role). The flow occurs by diffusion of water vapour from the intercellular spaces of the leaves (TARDIEU; DAVIES, 1992; TARDIEU; SIMONNEAU, 1998). Water flows through a pathway composed of a system of hydraulic resistances from the soil, passing through the plant and finally reaching the atmosphere (ZIMMERMANN; MEINZER; BENTRUP, 1995).

Several studies have described details of the resistance of each part of the soil-plant-atmosphere system, for example Angelocci (2002), Tuzet, Perrier and Leuning, (2003), Raats (2007) and Jong Van Lier et al. (2008). For water to be absorbed by the root, it must first overcome the hydraulic resistance of the soil itself, this resistance depending on the hydraulic properties of the soil, the water content and the distance to be travelled. Once the water is absorbed by the roots and reaches the xylem vessels, it encounters a low hydraulic resistance (DURIGON, 2011). From the xylem, the water rises inside the xylem vessels, depositing itself on the walls of the mesophyll cells, still as a liquid, after which the water evaporates, and is diffused, in the form of water vapour, into the intercellular spaces of the leaves until it reaches the atmosphere through the epidermis and cuticle, and/or through stomata; the cuticular route has a high hydraulic resistance, so stomata are the main diffusion route for water vapour from the leaves to the atmosphere (DURIGON, 2011).

The resistance of the air to the diffusion of water vapour in the vicinity of the leaves is represented by the resistance of the boundary layer or aerodynamic resistance. There is no consensus on which part of the system is responsible for the occurrence of water stress. Frequently, the reduction in soil moisture and the resulting increase in soil hydraulic resistance are indicated as the main mechanisms leading to water stress in plants (CARBON, 1973; HULUGALLE; WILLATT, 1983; SCHRODER et al., 2008). Water absorption by the roots depends on the movement of water in the soil towards the root surface but can also be influenced by transpiration. However, the rate of transpiration depends on stomatal conductance and the environmental conditions that affect transpiration (CAMPBELL; NORMAN, 1998). Carbon dioxide, oxygen and water vapour flow through the stomata, and in most plants the stomata remain open during the day and closed at night and under conditions of severe water stress. According to Pereira et al. (1997) water stress occurs in two situations: when the soil does not contain water available to the plants; when the soil contains available water but the plant is unable to absorb it at a sufficient rate and quantity to meet atmospheric demand (evaporative power of the air).

Water evaporation is a physical phenomenon that causes water to change state from the liquid phase to the gaseous phase directly from a liquid surface such as the sea, lake, river, etc., or from a moist surface such as plants and soil (BERLATO E MOLION, 1981). According to Philip (1957) the evaporation of water from the profile and surface layers of the soil can be divided into phases. In phase 1, the soil dries at a constant rate that depends only on the energy available at its surface, and is influenced by atmospheric demand, soil depth and its hydraulic properties; in phase 2 or the recession (depletion) phase, the surface dries, and evaporation takes place inside the soil. The water vapour reaches the surface by molecular diffusion and the mass flow is caused by fluctuations in air pressure; in phase 3, the evaporation rate in low humidity conditions loses its linearity and water moves through the profile as a result of adsorption forces between the water and the solid soil particles. In short, evaporation depends on the physical properties of the soil, which slowly transmits water to the surface to meet the demand induced by atmospheric conditions.

When this change in the physical state of water occurs in plants, it is called transpiration. Transpiration consists of the vaporisation of liquid water contained in the plant's tissues and the removal of the vapour into the atmosphere. This evaporation takes place through the stomata, which are microscopic structures (<50 pm) that occur on the leaves and allow communication between the inside of the plant and the atmosphere (PEREIRA et al., 1997). Practically all the water absorbed is lost through transpiration and only a small fraction is used inside the plant (COUTO and SANS, 2002).

The processes of evaporation and transpiration occur simultaneously on a vegetated surface. Evapotranspiration is the term used by Thornthwaith in the early 1940s to express this simultaneous occurrence (PEREIRA et al., 1997).

Thornthwaite (1948) defined potential evapotranspiration (Etp) as the amount of water used by a large area of actively growing vegetation under optimum soil moisture conditions. Reference evapotranspiration (Eto) was defined by Doorenbos and Pruitt (1977) as the water used by an extensive area of grass, actively growing, with a height of 0.08 to 0.15 m, completely covering the soil and without water deficiency. This definition of Eto coincides with that of Etp proposed by Thornthwaite (1948).

Jensen (1973) proposed alfalfa as a reference crop and defined reference evapotranspiration as that which occurs in an area without water deficiency, with a minimum border of 100 m planted with this crop, and the alfalfa must be 30 to 50 cm high. Actual evapotranspiration (Etr) is that which occurs on a vegetated surface, regardless of its area and soil moisture conditions (Thornthwaite, 1948; Pruitt et al., 1972; Villa Nova and Reichardt, 1989; Pereira, 1992).

Penman (1956) defined potential evapotranspiration as "the amount of water used in a unit of time by a low-growing, green crop, completely covering the surface, of uniform height and without water deficiency". Low-growing crops are implicit in this definition; however, Penman (1956) applied this concept only to low-growing grasses. The evapotranspiration of irrigated crops can be 10 to 30 per cent greater than that occurring on a grass surface.

The concept of crop evapotranspiration (Etc) was introduced by Doorenbos and Pruitt (1977), characterising it as the evapotranspiration of an agronomic crop, free of disease, growing on a cultivated area of one or more hectares, under optimum soil conditions, including water and fertility.

Evapotranspiration can be determined or estimated in different ways. According to Miranda et al. (2001), it can be measured using direct methods or estimated using climatic information. The first group includes the different types of lysimeters and the soil water balance, while the second includes theoretical and empirical methods such as Penman (1948), Thornthwaite (1948), Blaney and Criddle (1950), Jensen and Haise (1963), Priestley and Taylor (1972), Hargreaves (1977) and evaporimeters such as the "Class A" tank (SENTELHAS, 2003), among others.

Evapotranspiration can be measured using equipment called lysimeters or evapotranspirometers. Hillel et al. (1969), cited by Aboukhaled et al. (1986), defined lysimeters as containers that seek to represent the natural conditions of the soil-water-plant system, making it possible to regulate and conveniently control the processes that occur in the natural soil profile. Initially, lysimeters were used to study deep drainage and the concentration of nutrients extracted from the soil volume, and more recently they have been used to determine evapotranspiration and are widely used in agrometeorological research (PEREIRA et al., 1997).

According to Aboukhaled et al. (1986), lysimetry basically calculates the volumetric measurement of the water entering and leaving the system, which may or may not be covered by vegetation. The most commonly used lysimeters are drainage lysimeters, constant water table lysimeters and weighing lysimeters. Drainage lysimeters are based on the principle of conserving the mass of water in a volume of soil. This type of lysimeter works best over long periods of observation, around ten days, according to Camargo (1962) quoted by Pereira (2002).

Cunha et al. (1995) evaluated the effect of the climate on water consumption during the cycle phases of the IAC-165 rice-fallow cultivar in Jaboticabal, São Paulo, using deep drainage lysimeters of the "modified Thornthwaite" type to determine the maximum evapotranspiration of the crop. The Class A Tank, Solar Radiation, Penman, Linacre and Hargreaves methods were used to estimate reference evapotranspiration. The results showed that the ratio between maximum evapotranspiration and reference evapotranspiration increased as the plants developed, reaching maximum values during the flowering and grain filling subperiod, then decreasing until physiological maturity.

Carlesso et al. (2000) used a drainage lysimeter to assess changes in some morphological parameters of maize hybrids subjected to different irrigation water managements and to relate grain yield to the different irrigation rates applied, in the region of Santa Maria, state of Rio Grande do Sul. The results showed that irrigation rates greater than 30 mm reduce grain yield and dry mass accumulation, and that irrigation should be applied when the crop's maximum evapotranspiration indicates an accumulation of 20 to 25 mm.

A constant water table lysimeter consists of an asbestos cement box with a PVC pipe attached to one of its sides at the bottom and centre, which is connected to an intermediate tank via a hose. A cylindrical measuring tank with an outlet to the intermediate tank, with a socket for a tube with a graduated scale, provides the reading of the volume of water required by the evaporating surface (CURY and VILA NOVA, 1987).

Klosowski and Lunardi (2002) used water table lysimeters to determine the water consumption and crop coefficient of peppers grown in a protected environment. The result presented for total water consumption by the chilli crop over a 198-day cycle was 293.5 mm, with an average of 1.5 mm.day^{-1} . The highest water consumption was observed at the stage between flowering and the first harvest. The crop coefficient varied between 0.4 and 0.7 depending on the crop's stage of development.

Klosowski et al. (1999) used this same type of water table lysimeter to determine the water consumption and crop coefficient (kc) of Italian squash in the Botucatu region of the state of São Paulo. The results showed that total water consumption was 231.52 mm, with an average of 3.31 mm per day, for a 70-day cycle. The period of greatest water demand occurred between flowering and fruit development. The crop coefficient showed extreme values between 0.68 and 1.96, corresponding to the first and sixth weeks after planting.

Weighing lysimeters were classified by Aboukhaled (1982) into four types: mechanical; mechanical with electronic load cell; electronic load cell and hydraulic load cell. Weighing lysimeters with hydraulic load cells are devices that offer quality and precision results, as well as low cost and ease of construction and operation, with a resolution of up to 0.025 mm per reference evapotranspiration (ETo) reading (Freitas, 1994).

Hydraulic weighing lysimeters were described by Hanks and Shawcroft (1965), Tanner (1967) and Mcfarland et al. (1983). In Brazil, the first equipment of this type was built and described by Rodrigues (1987), for studies of grass evapotranspiration (ETo), in the city of Parnamirim, Pernambuco. Subsequently, other equipment of this type was built and evaluated in Brazil (Freitas, 1994). Different models and adaptations have been made to this equipment over the years. The main advantage of this equipment is its good precision on a daily scale and the low cost of construction compared to those built with electronic load cells, for example.

In the absence of direct measurements, such as those obtained using lysimeters, evapotranspiration can be estimated by indirect measurements using different methodologies, grouped according to Pereira et al. (1997) into five categories: empirical, aerodynamic, energy balance, combined and eddy correlations. Empirical methods such as the Class A tank, Thornthwaite, Camargo, Makink, solar radiation, Hargreaves-Samani, etc. are usually the result of correlations between evapotranspiration measured under standardised conditions and meteorological elements measured at standardised stations. The aerodynamic method is a micrometeorological method with a physical-theoretical basis in fluid dynamics and turbulent transport. The energy balance method accounts for the interactions of different types of energy with the surface. The combined methods portray the effects of the energy balance with those of the evaporative power. The eddy method is based on the vertical displacement of the atmosphere and the consequent transport of its properties. Indirect methods are characterised by the use of empirical equations or mathematical models, which require climatic-physiological data for their application. According to MELLO (1998), they present numerous precision problems, mainly due to the fact that they were developed for different climatic conditions to those in which they are normally applied.

Jensen etal . (1990) stated that in practice the estimated evapotranspiration of a specific crop (ETc) involves calculating the evapotranspiration of a reference crop (ETo), subsequently applying crop coefficients (Kc). Other ways of estimating reference evapotranspiration (ETo) are possible, for example, using evaporation from a free water surface. However, the rate of evaporation from tanks varies with the size of the tank and the boundary conditions. The same author suggests that ETo can be defined as "the rate at which water, if available, is removed from the soil surface and plants, from a specific crop, arbitrarily called the reference crop". ETo is usually expressed as latent heat rate per unit area or evaporated water blade. ETo is equivalent to potential evapotranspiration with the additional specification that it represents the evapotranspiration of a crop with ideal soil moisture and total coverage of the area. In order to manage water via the climate, it is necessary to know crop water consumption (ETc), which represents the amount of water that must be applied to the soil in order to maintain growth and productivity under ideal conditions (Pereira et al., 1997). According to Doorenbos and Kassam (1979), ETc is the result of the product of reference evapotranspiration (ETo) and the crop coefficient (Kc). Therefore, determining a crop's water consumption depends on knowing the reference evapotranspiration, which refers to the climatic conditions of the location where it is planted, and also the physiological and morphological characteristics that are peculiar to it, represented by its crop coefficient.

Sentelhas (2001) presents the Thornthwaite, Camargo, Hargreaves-Samani, Priestley-Taylor and Penman-Monteith methods as the most widely used, either due to their simplicity or degree of reliability. In addition to these, many other methods are presented in the literature, but due to their empirical or semi-empirical conditions, they are not as widely used as the majority of those mentioned, precisely because they were developed for specific climatic and agronomic conditions and are therefore not valid for different conditions.

ALLEN et al. (1998) comment that in May 1990, the FAO (Food and Agriculture Organisation of the United Nations) brought together several researchers from the International Commission on Irrigation and Drainage and the World Meteorological Organisation to review the methodologies used to estimate crop water requirements. As a result, the leading experts on the subject recommended the Penman-Monteith method as the standard for estimating ETo. This method, called PM-FAO 56, was selected because it presents results that are very close to the evapotranspiration of grass in different locations, because it represents the physical conditions present in the process and because it incorporates both physiological and aerodynamic parameters. To estimate ETo using this method, daily, weekly, decadal or monthly air temperature, relative humidity, radiation and wind speed data are required.

Cury and Villa Nova (1989) determined the values of cabbage crop coefficients obtained from various methods of estimating reference evapotranspiration and data measured by a water table evapotranspirometer. The results showed that a single annual crop coefficient should not be used, and that for greater accuracy of the results the appropriate coefficient should be used for each method of estimating reference evapotranspiration.

The crop coefficient used to estimate ETc, resulting from the ratio between ETo and ETc, considers ideal conditions, without local limitations for crop development, or reduced evapotranspiration due to soil water restrictions, planting density, diseases, spontaneous vegetation, insects or salinity. According to Pires et al. (2001), ETc differs from ETo mainly due to differences in soil cover, vegetative canopy properties and aerodynamic resistance between grass and crops. For the same crop, Kc varies depending on the stage of development and can reach a value greater than

1 when it is fully developed.

Other characteristics such as plant spacing, height, leaf area and foliage roughness also influence the variation in this coefficient. In addition to crop characteristics, Kc is intensely influenced by the humidity of the evaporating surface, and most Kc curves apply to crops well supplied with water (Sedyama et al, 1998). The effects of varying climatic conditions are incorporated into the ETo estimate. Therefore, Kc varies predominantly with the specific characteristics of the crop and the cultural practices adopted that affect development (Allen et al. 1998). This fact has justified the transfer of standard Kc values between locations and climates.

Doorenbos and Pruitt (1977) emphasise that it is essential to collect on-site data for irrigated crops, preferably on the vegetative period and crop development. The crop cycle is divided into four stages of development, which describe the following phenological periods: Kcini (beginning of crop establishment, number of days corresponding to the first 10% of the crop's vegetative development), Kcméd (full crop development, number of days corresponding to 70-80% of vegetative development) and Kcfinal (maturation period). Allen et al. (2006) also report that Kc varies according to the phenological stage of the crop and can reach, for example, a value greater than unity in the reproductive phase of many crops (Figure 1).

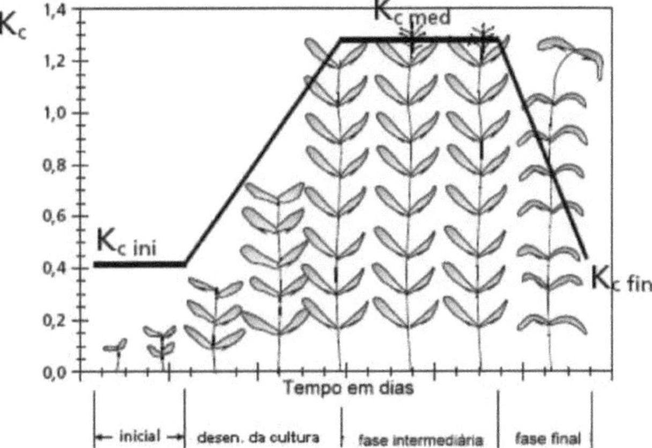

Figure 1. Generalised crop coefficient (Kc) curve (ALLEN et al., 2006).

2.3. Cultivation and evapotranspiration in a protected environment

Growing in a protected environment is a technique that minimises the effects of seasonal weather on crops and makes it possible to control temperature, relative humidity, oxygen levels, carbon dioxide levels, fertiliser levels and the amount of irrigation water applied. The advantages of growing in a protected environment include greater protection from climatic phenomena (frosts, excessive rainfall, a sharp drop in temperature during the night), protection of the soil from leaching, a reduction in fertiliser and pesticide costs and higher yields. According to Purquerio and Tivelli (2006), crop yields in a protected environment can be two to three times higher than those obtained in the open field and with superior quality. The use of a protected environment increases productivity and food production and requires less water, as evapotranspiration in this environment is around 60 to 80 per cent of that found in outdoor conditions (VAN DER POST et al., 1974; MONTERO et al., 1985; PRADOS, 1986; ROSENBERG et al., 1989). Martins (1992), using an umbrella-type greenhouse in two years of research, found that evaporation in a protected environment under plastic cover was around 30 per cent less than in the field, which contributed to lower crop evapotranspiration inside the protected environment. The FAO estimates that evapotranspiration is reduced by 30 per cent and water use per unit of production can be reduced by up to 50 per cent, resulting in higher crop yields in protected environments (STANGHELLINI, 1993). The evapotranspiration estimated by the Penman method in a protected environment with a polythene cover, with a density of 0.1mm, in Pelotas, RS, was between 45 and 70% of that found outdoors (FARIAS et al., 1994).

According to Montero et al. (1985), the lower evapotranspiration inside the protected environment is mainly due to the partial opacity of the plastic film to radiation and the reduction in the action of the winds, which interfere with the evaporative demand of the atmosphere, although the air temperature and relative humidity, at times, may be higher and lower respectively inside the protected environment than in the open, which would contribute to greater evapotranspiration.

2.4. Effect of water on IAF and chlorophyll content

The leaf area index (LAI) is the ratio of the plant's leaf area to the land area occupied by it. It is therefore possible to assess the growth and development of irrigated and rainfed crops based on this index, since a shortage or excess of water directly affects leaf development (MAGALHÃES, 1979).

The leaf area is represented by the plant's photosynthetically active surface and growth is related to plant production (TURNER, 1979). Numerous studies carried out on cultivated plants require knowledge of their leaf area at a certain stage or stages of their development. It is therefore related to plant metabolism, dry matter production and productivity (OLIVEIRA, 1977; SEVERINO et al. 2004).

There are various methods for determining leaf area, classified as destructive, non-destructive, direct or indirect. Direct or destructive methods mostly require detaching the leaves and are impractical in some studies (MARSHALL, 1968).

Oliveira (1977) proposed the linear dimensions method for determining leaf area (LA), according to which leaf area in bean plants can be estimated using the formula LA=k*(C*L), where

K=0.703 is a correction factor; C and L are the maximum length and width of the leaf, respectively. The leaf area index (LAI) can be obtained from the ratio between leaf area (LA) and the area of soil sampled. Monteiro et al. (2005), testing the dimension (C x L) and dry mass methods, concluded that the leaf area of the cotton plant can be estimated with good accuracy and excellent precision from measuring the dimensions of its leaves, with errors of around 10% and high correlation coefficients.

In sesame, Silva et al. (2002) found that estimates of leaf area are more accurate when both the length and width of the limbus are used. Lima et al. (2008), aiming to establish a model to estimate the leaf area of cowpeas, obtained the highest correlation coefficient value (R^2) for linear equations when using the product between the length and width (L x W) of the leaflets.

Chlorophyll, the main pigment responsible for capturing the light energy used in the process of photosynthesis, is one of the main factors related to the photosynthetic efficiency of plants and, consequently, their growth and adaptability to different environments. Chlorophyll absorbs all other wavelengths and reflects green wavelengths. Chlorophyll is a compound with a structure called porphyrin, which is a mixture of two substances: chlorophyll A (bluish green) and chlorophyll B (yellowish green). According to Taiz and Zeiger (2004), chlorophylls are located in the chloroplasts. This organelle is the site of photosynthesis, which has two important reactions: photochemical, in the membranes of the thylakoids, and biochemical, in the stroma of the chloroplast. In addition to chlorophylls, these organelles contain other pigments known as accessories, such as carotenoids (carotenes and xanthophylls).

Chlorophyll molecules have the ability to transform solar radiation into chemical energy through the process of photosynthesis (SANTOS and CARLESSO, 1998). Plants are the primary transformers of solar energy and their efficiency is a determining factor in agricultural productivity. Water deficit is one of the environmental stresses responsible for the loss of pigments in leaves, altering the plant's life cycle. In addition, the ratio between chlorophyll a and b in terrestrial plants can be used to indicate the response to shading and premature senescence, and the ratio between chlorophyll and carotenoids is used to a lesser extent to diagnose the rate of senescence under water stress (HENDRY and PRICE, 1993).

According to Engel and Poggiani (1991), photosynthetic efficiency is linked to the chlorophyll content of plants, affecting growth and influencing their adaptability to different environments. Chlorophyll, the main pigment responsible for capturing the light energy used in the photosynthesis process, is one of the main factors related to the photosynthetic efficiency of plants and consequently to their growth and adaptability to different environments.

Physiological parameters, such as the indirect measurement of chlorophyll content in leaves, can be used as a tool to diagnose the integrity of the photosynthetic apparatus when plants are subjected to environmental adversities, as they are quick, precise and non-destructive techniques

(VAN DEN BERG and PERKINS, 2004; TORRES NETTO et al, 2005).

2.5. The effect of water on crop yields

Most crops have critical periods in terms of water deficiency, during which the lack of water causes serious decreases in final production; the damage caused depends on its duration and severity and the plant's stage of development (FOLEGATTI et al., 1997). In order to obtain optimum production, the response of the water supply on yield must be known.

According to Doorenbos and Kassam (1994), this response is quantified by the crop response factor (ky), which relates the relative yield drop [1- (yr/ym)] to the relative evapotranspiration deficit [1-(ETr/ETm)]. Achieving high yields also requires the adoption of management practices aimed at improving irrigation and this can only be done with knowledge of the crop's water requirements. According to the aforementioned author, crops can be classified according to their sensitivity to water stress in four categories: low (Ky < 0.85); low/medium (0.85 < Ky < 1.00); medium/high (1.00 < Ky < 1.15) and high (Ky > 1.15). Cordeiro et al. (1998), studying sensitivity to water deficit in cowpea, found ky values of less than 1, indicating low sensitivity to the water deficits applied in the research. Doorenbos and Kassam (1994) studying phaseulus obtained ky = 1 at flowering, ky = 0.75 at grain filling and ky = 0.2 at the vegetative stage.

Research has been carried out to assess the sensitivity of the bean crop to water stress depending on its stage of development. According to Fageria et al. (1991), the stage of the plant most sensitive to water deficiency is the reproductive stage, which is highly vulnerable from the start of flowering to the start of pod formation. Productivity is most affected when water stress occurs between 5 and 10 days before anthesis, with yields falling by more than 50 per cent (NORMAN et al., 1995). This effect is mainly caused by the low pollination rate and the abortion of ovules, which cause the reproductive organs to abscond, resulting from a decrease in the translocation of photoassimilates from the leaves to the flowers (KRAMER; BOYER, 1995).

OLIVEIRA (1987) evaluated water deficit in the bean crop in a greenhouse and concluded that water deficit in the growth, flowering and fruiting phases caused a 31.2 per cent, 10.8 per cent and 51.8 per cent reduction in grain production, respectively. With reduced cell expansion, there is a reduction in the source (leaves) and, consequently, in the photosynthates available for translocation towards the grains, resulting in a reduction in the size of the demand. If the water deficit occurs after leaf expansion, there will be less competition between leaves and fruit for photosynthates, and the demand will only be affected by the lower availability of photosynthates, i.e. a decrease in the photosynthetic rate (CONFALONE et al., 1998). Water stress develops in the plant when the rate of transpiration exceeds the rate of absorption and transport of water in the plant (BERKOWITZ, 1998).

Considering the current worldwide concern about the scarcity of water resources and their high cost in certain situations, the search for increased efficiency in the use of water by crops has been a cause for concern for research, extension and rural producers, since this component of production increasingly occupies an important share of production costs. According to Reichardt and Timm (2004), around 98% of the volume of water absorbed by the plant is lost to the atmosphere through transpiration. However, this flow is necessary for the plant's development and therefore humidity limits must be kept within the optimum range for plants. Continuous water absorption is essential for growth and vegetative development, as most plants in tropical climates lose more than their weight in water per day under certain conditions (PIMENTEL, 1998).

According to Letey (1985), maintaining soil moisture within the optimum range, understood as the lower limit where mechanical resistance to root development begins to occur and the upper limit where low aeration occurs, would provide greater root development. Therefore, the amount of water applied combined with the physical properties of the soil influence water potential, aeration and mechanical resistance, which are directly related to production. In some regions where water is the main limiting factor, the aim should be to obtain maximum production per unit of water applied, adapting irrigation to critical periods of water deficit such as: germination, flowering and grain filling or fruit formation (BERNARDO, 1995). According to the author, the ratio between the water evapotranspirated by the crop and that applied by irrigation must be close to 1.0 in order to achieve maximum efficiency in the use and application of water.

Ertek et al. (2005) determined the irrigation water use efficiency (IWUE) and the plant evapotranspiration water use efficiency (WUE) for the cucumber crop. These authors, evaluating irrigation treatments consisting of two irrigation intervals (4 and 8 days) and three plant-tank coefficients (Kcp) (0.50, 0.75 and 1.0), obtained the highest IWUE (0.089 $t.ha^{-1}$ .mm) and WUE (0.079 $t.ha^{-1}$ .mm) for the treatment with an irrigation interval of 8 days and Kcp equal to 1.0.

Antony and Singandhupe (2004), evaluating the influence of drip and surface irrigation on growth, yield and water use efficiency (WUE) of the pepper crop (Capsicum annuum L.) var. California Wonder, concluded that WUE had a significant negative linear relationship with net photosynthesis, i.e. when photosynthesis rates increased, WUE decreased linearly.

At lower levels of irrigation, there is less water and consequently the stomata close, causing a reduction in water loss and a decrease in CO_2 fixation. Countries with less water or that don't want to or can't afford this type of production prefer to buy grain abroad, transferring the environmental cost to the producing countries. It is in this sense that it is said that "exporting grain is exporting water" (BROWN, 2003).

CHAPTER 3

MATERIAL AND METHODS

3.1 Experimental procedures

The experiment was carried out in a greenhouse with a 15 *pm* thick low density polythene cover in the experimental area of the Teresina Agricultural College of the Federal University of Piauí (UFPI), in the municipality of Teresina-PI, at coordinates 05°05'21" south latitude and 42°48'07" west longitude and altitude 74 metres above sea level. According to the climate classification of Thornthwaite & Mather (1955), the region's climate is C1sA'a', characterised as dry subhumid, megathermic, with a moderate water surplus in summer and a concentration of 32.2% of potential evapotranspiration in the September - October - November quarter, with average annual rainfall of 1500mm, concentrated between the months of January to May, an average temperature of 27°C and average relative humidity of 74% (ANDRADE JÚNIOR et al., 2005).

The crop was grown in pots, with soil classified as ARGISSOLO VERMELHO-AMARELO, Distrófico, textura franco-arenosa; very deep, acidic, with flat relief, collected in the 0 to 30 cm layer, and samples were also taken for its physical-chemical characterisation, as shown in Table 1. To standardise the filling of the pots, in order to obtain soil with the same density as in the field, its density was determined using the volumetric ring method and its moisture content. In this way, the mass of soil to be placed in each pot was determined, considering that it had a volume of eight litres.

Table 1. Physico-chemical characterisation of the soil used in the pots, Teresina-PI, 2012.

Layer	PH	MO	P	K	Ca	Mg	In	Al	H+Al	S	CTC	V
Cm	H_2O	$g.kg^{-1}$	--mg dm^{-3} --				---------------------cmolc dm^{-3} ----------------					%
0-10	5,3	6,8	8,5	0,14	2,1	0,8	0,03	0,04	1,5	3,1	4,5	66,5
10-20	5,4	2,6	10,0	0,10	1,7	0,5	0,03	0,00	1,4	2,3	3,7	62,4

Source: UFPI Soil Analysis Laboratory

Bean seeds with a determinate growth habit, semi-erect "bush" type and uniform maturity were used, coming from the UFPI germplasm bank. Foundation fertiliser was applied to the pots, using NPK mixed fertiliser (5-30-15) to apply 10 kg of N, 60 kg of P_2O_5 and 30 kg of K_2O ha^{-1}. Planting took place on 10 March 2012, with four seeds being sown per pot. Ten days after emergence, thinning was carried out, leaving two plants per pot. At 20 days after emergence, top dressing was applied with 40 kg of N and 100 kg of K_2O ha^{-1}. During the experiment, phytosanitary control was carried out with an application of the fungicide metalaxyl-m + mancozeb at 10 days after emergence.

The pods were harvested when they appeared to be ripening in the field. The pods were harvested every week starting at 80 days after emergence. After harvesting, the pods were packed in paper bags and then counted and weighed. After the harvest was completed at 120 DAE, the grain moisture content was corrected to 13% in order to obtain the grain yield.

3.2. Crop evaporation and evapotranspiration.

Twelve drainage lysimeters were used to determine crop evaporation and evapotranspiration. The lysimeters were made from plastic pots with a capacity of 8 litres, which were perforated at the base and a 10 cm long hose with an internal diameter of one inch was fitted into this hole, connecting it to a plastic container with a capacity of 0.5 litres to control and collect the drainage water.

A non-woven geotextile blanket (Bidim OP 30) was placed at the bottom of each pot to prevent soil loss during drainage. A 3 cm layer of gravel and another covering of geotextile blanket was placed on top of this to facilitate water drainage. All the pots were then filled with soil.

Four lysimeters were planted with fava beans to determine crop evapotranspiration and four were left with only soil to determine soil evaporation without cultivation.

Evaporation and evapotranspiration of the fava bean crop were determined using the lysimetry method, which consists of a water balance based on the law of conservation of mass, presented by Reichardt (1985) equation 1:

$$P + I - D - E = \pm h \tag{01}$$

In which:

P: natural precipitation, in mm

I: irrigation depth, in mm

D: drainage slope, in mm

E: crop evaporation or evapotranspiration, in mm

h: variation in soil water storage within the lysimeters, in mm.

Considering that the application of irrigation always raised the humidity in all the lysimeters to field capacity, the variation in storage is equal to zero, and also because of the plastic covering of the greenhouse, rainfall was disregarded. So the equation for calculating crop evaporation and evapotranspiration was reduced to the following expression.

$$E = I - D \tag{02}$$

The volume of water from evaporation and crop evapotranspiration, respectively, were determined daily in each pot, obtained by the volume applied in each pot minus the respective volume drained the following day. To transform the volumes of evaporation, reference evapotranspiration and crop evapotranspiration obtained in each pot into blade values, they were divided by the average area of the pots.

Irrigation was carried out daily at a rate corresponding to 100 per cent of crop evapotranspiration in the treatments without water stress, while in the treatments under water stress irrigation was carried out at a rate corresponding to 50 per cent of crop evapotranspiration (Figure 2).

Figure 2: Irrigation details for the treatments

3.3 Treatments and experimental design

The treatments consisted of a combination of inducing water stress of 50 per cent of the fava bean's evapotranspiration by crop development stage (Figure 3). The fava bean crop cycle was divided into four phases (I, II, III, IV): phase I, from emergence to 20 days after emergence (DAE); phase II, from 20 to 40 DAE; phase III, 40 to 60 DAE and phase IV, 60 to 100 DAE.

The experiment used 10 treatments with four replications (pots) in a completely randomised design (DIC), making up 40 plots, each pot containing two plants. The treatments were: Treatment 1: Water stress in the vegetative phase (I); Treatment 2: Water stress in the vegetative phase (II); Treatment 3: Water stress in the reproductive phase (III); Treatment 4: Water stress in the reproductive phase (IV); Treatment 5: Water stress in the vegetative phases (I and II); Treatment 6: Water stress in the vegetative and reproductive phases (II, III and IV); Treatment 7: Water stress in the reproductive phases (III and IV); Treatment 8: Water stress in the vegetative and reproductive phases (I, II and III); Treatment 9: Water stress in the vegetative and reproductive phases (I, II, III and IV) and Treatment 10: Full irrigation.

The ASSISTAT software Version 7.6 beta (SILVA et. al., 2009) was used for statistical analysis. To interpret the results, analysis of variance was used, applying the "F" test and, when significant, the Tukey test was applied to rank the means of the treatments.

Figure 3: Detail of the experimental area with developing fava beans

3.4 Variables assessed

The variables assessed were: number of pods per plant (NV), number of grains per pod (NG), grain weight (PG), leaf area index (IAF) and dry matter (MST) and yield of pods and grains: the number of flowers emitted each week and the number of pods formed were quantified on each fava bean plant (a pod was considered when at least one seed suitable for germination was present) and, by difference, the number of aborted flowers and pods was obtained.

The chlorophyll content was determined at 40, 48, 60 and 80 days after emergence, always taking the reading on the leaves in the middle third of the plants, with three repeat readings in each plot (Figure 4). To calculate the chlorophyll content, we used the portable chlorophyll meter, chlorofiLOG, which allows instantaneous readings of the relative chlorophyll content in the leaf without destroying it, making this method simple to operate and quick to measure; it correlates well with values obtained in laboratories, as well as allowing a non-destructive assessment of leaf tissue (FALKER, 2008).

Figure 4: Details of chlorophyll content determination

The chlorophyll meter evaluates two points: one of high absorbance, in the region of the red spectrum, where there is a peak of absorbance by chlorophyll, and another in the region of the infrared spectrum, where there is maximum transmittance; the latter to remove the effect of the thickness of the leaf and its degree of hydration. Thus, the chlorophyll content can be estimated indirectly, as the device measures the intensity of the green colour of the leaf, which is proportional to the chlorophyll concentration, which in turn is related to the leaf nitrogen content (BUZETTI, et al., 2008).The leaf area index was calculated at 40 days after emergence, at the beginning of flowering. The leaf area (LA) was determined using the linear dimensions method proposed by Oliveira (1977), according to which the LA leaf area in bean plants can be estimated by multiplying the measurements of the length and width of the leaflets multiplied by a correction factor equation (03):

$$AF = K(C * L) \tag{03}$$

Being:

K=0.703 (correction factor);

C= Leaf length;

L= maximum leaf width.

The leaf area index (LAI) for each treatment was obtained from the ratio between leaf area (LA) and the area of soil sampled (ΔS) equation (04).

$$IAF = \frac{AF}{\Delta S} \tag{04}$$

The maximum length and width of the leaflets were obtained using a millimetre ruler. To estimate leaf area, the number of leaves per plot was quantified and the length and width of a sample of 10 leaflets in each plot were measured.

To determine the total dry matter at the end of the crop cycle, the area of the plant and the root system were harvested and taken to the oven (65°C) for 48 hours and then weighed: abortion index, as a percentage (the total number of flowers and pods aborted, in relation to the total number of flowers emitted); number of pods per plant (total number of pods in each pot divided by the number of plants per pot); number of grains per pod (dividing the number of grains by the number of pods); grain yield (obtained by weighing the grains of the plants in each pot, correcting this weight to 13% humidity).

3.5 Crop response factor to water deficit (Ky)

To quantify the effects of water stress on the various phenological stages of the plant, the crop's sensitivity to water deficit (Ky), the empirical expression described by Doorenbos and Kassam (1994) was used, which quantifies the relationship between the reduction in relative yield and the evapotranspiration deficit.

$$Ky = \frac{[1-(yr/ym)]}{[1-(ETr/ETm)]} \tag{05}$$

Being:

ky - partial sensitivity factor for each stage or phenological phase of the fava bean yr - actual crop yield obtained in the treatments subjected to water stress;

ym - maximum crop yield obtained in the treatment that did not suffer water stress;

ETr - actual crop evapotranspiration obtained in the treatments subjected to water stress;

ETm - maximum crop evapotranspiration obtained in the treatment without stress;

The potential yield (Ym) and crop evapotranspiration (ETm) were obtained from the treatment corresponding to 100% replacement of the water consumed.

3.6 . Water use efficiency (EUA)

To determine water use efficiency (WUE), the methodology described by Doorenbos and Kassam (1979) was used, in which the water use efficiency (WUE) of crops can be determined both for biological production and for the production of the whole plant or part of it. In this work it was determined for dry matter production, as well as pod production and dry grain production of fava beans, using the ratio between dry matter produced, in kg ha^{-1} , and the amount of water consumed by the crop in the plot in m^3 ha^{-1} , expressed in kg m^{-3} . Water use efficiency was obtained using equation 5:

$$\mathbf{EUA} = \frac{Y}{W} \tag{06}$$

In which:

EUA: water use efficiency, (kg m^{-3} ;

Y: crop yield, (kg ha^{-1} ;

W: Amount of water consumed by the crop, (m^3 ha^{-1}

CHAPTER 4

RESULTS AND DISCUSSION

4.1. Water deficit and fava bean phenology

The fava bean seedlings emerged on average eight days after sowing. At this stage, all treatments received a water table equivalent to evapotranspiration and were not subjected to water deficit.

The analysis of variance for the start of flowering, as well as the start of ripening and the fava bean cycle, in days after emergence (DAE), are shown in Table 2. There was no significant difference between the treatments in terms of the start of flowering, which occurred on average at 36 DAE. Although there was no significant difference between the treatments, those under water stress only in the vegetative phase (I) and only in phase (II) flowered at 34 DAE, while the treatment with full irrigation flowered at 37 days and the treatment with water stress in phases II, III and IV flowered at 40 days, indicating a slight induction of precocity due to water stress in the vegetative phase (Table 3). The time taken for flowering to start is also a characteristic of the variety. Santos et. al. (2002) observed differences of up to 20 days in the start of flowering of fava beans between 8 varieties studied, with the earliest starting flowering at 49 days and the latest at 71 days after sowing. Silva Neto (2010) studied 70 fava bean accessions and found that the average flowering time was 65 days, with a range of 59 days. In the same study, this author observed that the earliest genotypes reached full flowering at 39 days, while the latest at 98 days after emergence.

The analysis of variance showed that there was no significant difference between the treatments when it came to the onset of pod ripening, with the onset of pod ripening occurring at 74 DAE (Table 3). In the treatments E.H. in phase (I), E.H. in phase (II), E.H. in phases (I, II and III), the start of ripening was at 70, 70 and 72 days after emergence respectively, while in the treatment with full irrigation the start of ripening was at 76 DAE and in the other treatments; E.H. in phases (II, III and IV); E.H. in phases (I and II) and E.H. in phase (III) at 76 DAE. H. in phase (III) at 76 DAE. The ripening of the fava bean occurred unevenly, and even in the ripening phase, some branches continued to emit inflorescences, especially in the treatments that were receiving the amount of water required by the crop during the ripening phase (Treatments without water deficit in phase IV).

As for the analysis of variance for the crop cycle (DAE), there was no significant difference by the F test, and the crop cycle averaged 117 days. The treatments, E. H. in phase (I), E. H. in phase (II), E. H. in phase (III) had a cycle of 120 DAE, showing that after a period of water deficit with the return of full irrigation the plants had a longer vegetative period, because in the treatments with full irrigation, E. H. in phase (IV), E. H. in phases (III and IV) and E. H. in phases (I, II, III and IV) the plants had a longer vegetative period. H. in phases (I, II, III and IV) had a cycle of 115, 115, 112 and

111 DAE, respectively. Oliveira et. al., (2010) studying the phenology and vegetative development of fava beans, the fruit ripening phase began between 51 and 74 days, the maturity of 50% of the pods between 67 and 85 days and complete ripening occurred between 84 and 97 days, as for the cycle, under greenhouse conditions, the accessions tested had a cycle of 120 days, with field ripening occurring between 84 and 97 days.

Table 2. Summary of analyses of variance for days to flowering, maturity and cycle (DAE) of fava beans, under water stress by development stage and full irrigation.

F. V. -	Flowering (DAE)	Maturation (DAE)	Cycle (DAE)
	QM	QM	QM
Treatments	16.6ns	25.06ns	49.55ns
Waste	11,81	13,33	16,55

ns not significant (p >= .05)

Table 3. Duration values, in days after emergence (DAE), for flowering, pod ripening; crop cycle under water stress and full irrigation.

TREATMENTS	Flowering	Ripening	Cycle
	AED	AED	AED
E. H. in phase (I)	34 a	70,00 a	120,00 a
E. H. in phase (II)	34 a	70,00 a	120,00 a
E. H. in phase (III)	37 a	76,00 a	120,00 a
E. H. in phase (IV)	36 a	72,00 a	115,00 a
E. H. in phases (I and II)	36 a	76,00 a	120,00 a
E. H. in phases (II, III and IV)	40 a	76,00 a	120,00 a
E. H. in phases (III and IV)	35 a	72,00 a	112,50 a
E. H. in phases (I, II and III)	35 a	72,00 a	119,50 a
E. H. in phases (I, II, III and IV)	37 a	74,00 a	111,25 a
With full irrigation	37 a	76,00 a	115,00 a
	CV=9.25%	CV=4.97%	CV=3.47%
	MG=36,00	MG=73.4	MG=117.32

* Averages followed by the same letter do not differ according to Tukey's test at a 0.05 probability level.

The summary of the analysis of variance for the leaf area index of the fava bean can be seen in Table 4. There was a difference in the LAI, at 5% by the F test, for water stress in the stages of fava bean development. The fava bean had the lowest IAF in the treatments water stress in vegetative phase II and water stress in phases II, III and IV, with IAFs of 1.69 and 1.39 respectively. In the treatment without water deficit, with full irrigation, the IAF was 3.20 (Table 5). The reduction in leaf area in plants under water deficit may be a survival strategy aimed at reducing the area available for transpiration (Correia and Nogueira, 2004). This reduction is a morphological defence mechanism, as the reduction in the interface between the plant and the atmosphere reduces transpiration, which is positive, but also reduces photosynthetic assimilation, which is negative for production. With a smaller leaf area, there is a reduction in transpiration, conserving water in the soil for a longer period (TAIZ and ZEIGER, 2004). According to Fernández et al. (1996), the use of water by plants is determined by leaf area and, once exposed to water deficit, this is reduced. According to Taiz and Zeiger (2004), there is a close relationship between the availability of water in the soil and leaf area, with leaf growth decreasing as soil moisture decreases, suggesting that this variable is highly sensitive to water deficiency.

The summary of the analysis of variance for the flower and pod abortion index is shown in Table 4. There was a significant effect of water deficit on the flower and pod abortion index at the 1% level. The full irrigation treatment had a lower abortion rate than all the others, with a rate of 27.10%. The other treatments did not differ, with values of 73.05; 67.35; 64.88 and 62.86 % when the deficit occurred in phases I, II and III; I and II; III and IV; II, III and IV, respectively. Water stress during the flowering and pod formation period intensified the abortion of fava bean flowers and pods. Similar results were found by Hostalácio and Válio (1984) where water stress during the flowering and pod formation period of bean plants induced the abortion of flowers and young pods, helping the formation of the first ones, because there was a source-drain competition, so abnormal flowers were eliminated, with fertilisation failure, or younger pods that aborted due to a lack of nitrogen or carbohydrates. Water stress during the pod filling stage led to young pods aborting and the production of shrivelled pods.

Table 4 Summary of the analysis of variance for leaf area index of fava beans at 40 DAE; flower and pod abortion index of fava beans under water stress by development stage and full irrigation.

| F. V - | Leaf area index | Flower and pod abortion rate |
	QM	QM
Treatments	0,541 *	858,5 **
Waste	0,223	145,4

* significant at 5% probability level (.01 =< p < .05)
** significant at 1% probability level (p < .01)

Table 5: Leaf area index; flower and pod abortion index of fava beans under water stress and full irrigation.

TREATMENTS	Leaf area index	Abortion Index %
E. H. in phase (I)	2.04 ab	59.94 ab
E. H. in phase (II)	1,69 b	58.34 ab
E. H. in phase (III)	1.90 ab	61.43 ab
E. H. in phase (IV)	2.19 ab	59.17 ab
E. H. in phases (I and II)	2.62 ab	67,35 a
E. H. in phases (II, III and IV)	1,39 b	62.83 ab
E. H. in phases (III and IV)	2.63 ab	64.88 ab
E. H. in phases (I, II and III)	2.08 ab	73,05 a
E. H. in phases (I, II, III and IV)	2.08 ab	64.68 ab
With full irrigation	3,20 a	27,10 b
	CV=26.93%	CV=26.30%
	MG=1.76	MG=59.88

Table 6 shows the summaries of the analyses of variance for the chlorophyll content index at 40, 48, 60 and 80 days after emergence in the fava bean crop, as a function of water stress. It can be seen that there was a significant effect of water stress on chlorophyll content at 40 and 48 DAE, with plants that suffered water stress in phase III having lower chlorophyll contents than non-stressed plants. Water stress interferes with the plant's nutritional status, reducing the ICF (Table 7). These results differ from those obtained by Lima (2008) with the bean crop, who did not record significant variations in ICF that could be correlated with the effects of water deficiency. At 60 and 80 DAE (Table 7), there was no statistically significant difference in chlorophyll content between the treatments. The chlorophyll content in the treatment with full irrigation was 39.05 at 60 DAE and 35.37 at 80 DAE; in the treatment with water stress in phases I, II, III and IV, the chlorophyll content was 31.52 at 60 DAE and 29.46 at 80 DAE; in the phase I water stress treatment, the chlorophyll content was 36.25 at 60 DAE and 33.07 at 80 DAE; in the phase II water stress treatment, the chlorophyll content was 35.35 at 60 DAE and 33.75 at 80 DAE; in the other treatments, the same decrease in chlorophyll content occurred from 60 to 80 DAE. At 60 and 80 DAE, the plants are already in the pod ripening phase and senescence is beginning, with leaf fall and a decrease in chlorophyll content in all treatments (Figure 5).

Table 6. Summary of analyses of variance for ICF chlorophyll content in fava beans at 40, 48, 60 and 80 DAE, under water stress by development stage and full irrigation.

F. V.	Chlorophyll content (ICF)			
	40 (DAE)	48 (DAE)	60 (DAE)	80 (DAE)
	QM	QM	QM	QM
Treatments	44,70*	35,71 *	22.30 ns	10.30 ns
Waste	15,19	11,72	48,37	21,49

* significant at 5% probability level (.01 =< p < .05) ns not significant (p >= .05)

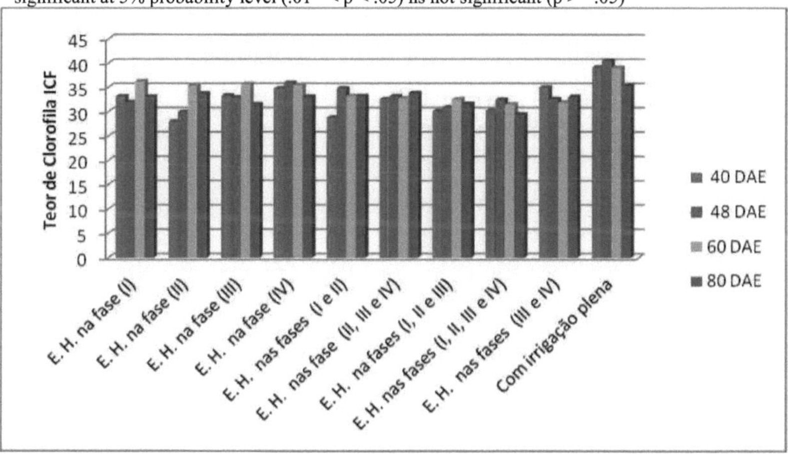

Figura 5.   Average chlorophyll content in fava beans under water stress at 40, 48, 60 and 80 days after emergence.

Table 7. Average values of the chlorophyll content index (CCI), measured at 40, 48, 60 and 80 DAE, in water-stressed and unstressed fava beans by development stage.

TREATMENTS	Chlorophyll content ICF			
	40 DAE	48 DAE	60 DAE	80 DAE

E. H. in phase (I)	33.15 ab	31,97 b	36,25 a	33,07 a
E. H. in phase (II)	28.00 ab	29,97 b	35,35 a	33,75 a
E. H. in phase (III)	33.32 ab	32.97 ab	35,67 a	31,55 a
E. H. in phase (IV)	34.77 ab	35.95 ab	35,45 a	33,07 a
E. H. in phases (I and II)	28,82 b	34.80 ab	33,24 a	33,25 a
E. H. in phases (II, III and IV)	32.60 ab	33.17 ab	32,80 a	33,85 a
E. H. in phases (III and IV)	35.00 ab	32.60 ab	31,92 a	33,07 a
E. H. in phases (I, II and III)	30.17 ab	30,85 b	32,60 a	31,61 a
E. H. in phases (I, II, III and IV)	30.22 ab	32.47 ab	31,52 a	29,46 a
With full irrigation	39,12 a	40,47 a	39,05 a	35,37 a
	CV=11.99%	CV=10.21%	CV=20.23	CV=14.13%
	MG=32.5	MG=33.5	MG=34.38	MG=32.80

* Averages followed by the same letter in the column do not differ according to Tukey's test at a 0.05 probability level.

4.   2Water-soil-plant-atmosphere relationship in a protected environment

The average daily evaporation values inside the greenhouse are shown in Figure 6. In this study, the average daily evaporation from the surface of the moist soil during the experiment was 1.10 mm per day^{-1} with a standard deviation of 0.18 mm per day^{-1} , since evaporation takes place in a protected environment, the atmospheric demand is reduced, thus reducing the rate of evaporation from the soil. This is due to the fact that wind speed, air humidity, air temperature and incident solar radiation are higher outside the greenhouse than inside (VÁSQUEZ, et al., 2005). The same author, studying the effect of the protected environment on meteorological elements when growing melons, found average values for internal relative humidity (IRH) and external relative humidity (ERH) during the cycle of 71.17% and 74.25%, respectively, so that the average IRH was 3.08% lower than outside.

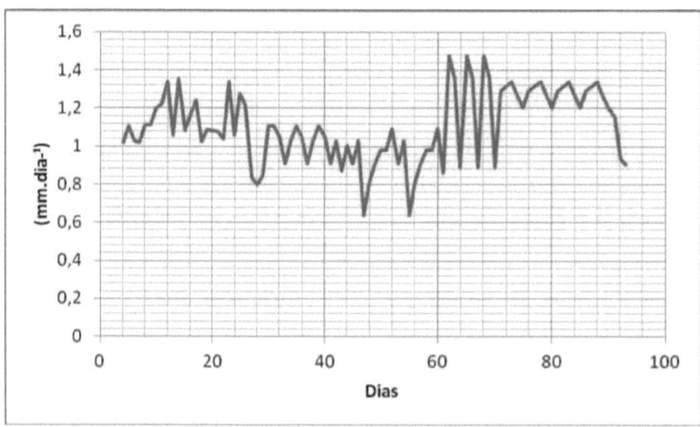

Figura 6.   Average values of daily evaporation from the soil surface in a protected environment during the experiment.

The transpiration of the fava bean crop in a protected environment varied from 0.37 mm.day^{-1} during the initial stage of vegetative development (from emergence to 20 DAE), reaching 2.43 mm.day^{-1} at the start of flowering at 40 DAE and reaching maximum transpiration at the grain filling stage of 3.65 mm.day^{-1} from 40 to 74 DAE, at the end of the cycle the average transpiration was 0.65 mm.day^{-1} Figure 7. OLIVEIRA et. al. (2005) studying indicators of water stress in bean plants found average transpiration values throughout the cycle between 2 and 12 mmol m^{-2} s^{-1} , with maximum values obtained at flowering where transpiration reached maximum values of 12 mmol m^{-2} s^{-1} .

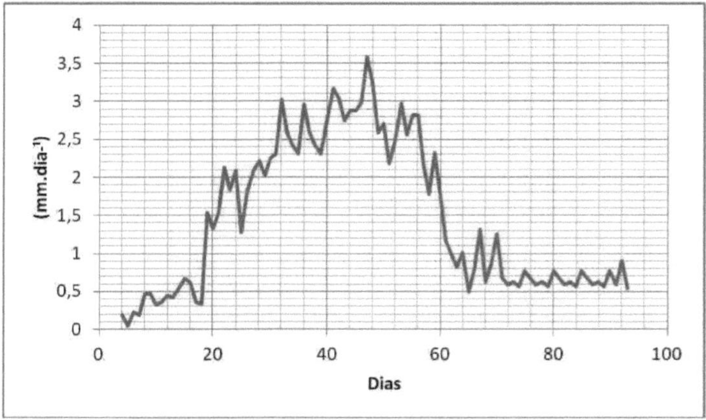

Figura 7.   Average daily transpiration values of fava beans from emergence to the end of the cycle in a protected environment.

The evapotranspiration of the fava bean crop in a protected environment varied from 1.51 mm.day^{-1} during the initial stage of vegetative development (from emergence to 20 DAE), reaching 3.12 mm.day^{-1} at the start of flowering at 40 DAE and reaching the maximum evapotranspiration at the grain filling stage of 3.75 mm.day^{-1} , from 40 to 74 DAE, in the final phase of grain filling and physiological maturation of the pods when the leaves begin to senesce at the end of the crop cycle, the average evapotranspiration was 1.90 mm.day^{-1} , with a total evapotranspiration of 228.66 mm during the cycle Figure 8. In average terms, during the whole of the fava bean cycle its water requirement, depending on the variety and local soil and climate conditions, can exceed 700 mm (MOUSINHO, 2010). Ritter and Scarbough, cited by Mousinho, (2010), when working with fava beans on the east coast of the United States under field conditions, found daily evapotranspiration values ranging from 6.4 to 9.1 mm. Cunha (2001), studying agrometeorological parameters in the cultivation of peppers in field and protected environments, found evapotranspiration values for protected cultivation of 509.60mm, and a value of 741.11mm for field cultivation, thus showing a percentage increase of 45.43% from protected to field cultivation, This may be associated with the partial opacity of the plastic film to radiation and the reduction in the action of the winds in this crop, which are the main factors in the evaporative demand of the atmosphere. Similar results were also found by Farias et al. (1994) who studied evapotranspiration estimated by the Penman method in a protected environment with a 0.1mm density polythene cover in Pelotas, RS, finding evapotranspiration values between 45 and 70% lower in the protected environment than in the field.

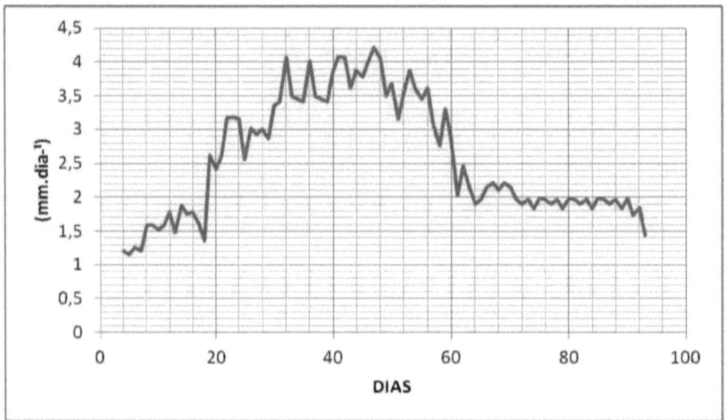

Figura 8.   Average daily evapotranspiration values of fava beans from emergence to the end of the cycle in a protected environment.

4.3 . Effect of water deficit on the productive performance of fava beans

The summary of the analysis of variance for the mass of 100 grains showed that there was no significant difference between the treatments, water stress by stage of development and full irrigation (Table 8). The average 100-seed mass among the treatments was 31.8 g (Table 9). The values for the mass of 100 grains per treatment were E. H. in stage (III), E. H. in stage (I) and E. H. in stages (III and IV). H. in phases (III and IV) with values of 36.61; 36.55 and 33.79 grams per 100-grain mass respectively, while in the treatments with full irrigation, E. H. in phases (I and II), E. H. in phases (I, II, III and IV), E. H. in phases (I, II and III) and E. H. in phase (II) with values of 26.73; 25.60; 29.60; 30.59 and 30.24 grams per 100 grains respectively.

The results obtained were below the range found by Santos et al. (2002), which was 32.6 to 79.5g (study of 8 varieties of fava beans) and between the range found by Melo (2005) who found a variation of 30.96 to 82.29 g (trial with 13 varieties of fava beans). Azevedo et al. (2003) studied a variety of fava bean and observed an average variation in the weight of 100 fava bean

34

seeds from 47.39 to 90.05 g.

The analysis of variance revealed that the number of pods per plant showed a significant difference (P < 0.01) by the F test (Table 7), for water stress by stage of development and full irrigation. The highest number of pods per plant was found in the treatments with full irrigation, E.H. in phase (II), E.H. in phase (I), E.H. in phases (I and II) and E.H. in phase (IV). H. in phase (IV); with 13.50; 9.62; 6.37; 6.12 and 6.00 number of pods per plant respectively. The lowest number of pods per plant was found in the treatments E. H. in phases (I, II and III), E. H. in phases (I, II, III and IV), E. H. in phase (III), E. H. in phases (II, III and IV) and E. H. in phases (III and IV). H. in phases (III and IV) with 2.5; 3.12; 3.75; 4.0 and 4.37 number of pods per plant respectively, it should be noted that the decrease in the number of pods was due to the higher rate of flower and pod abortion, when the water stress occurred at the beginning of flowering and ripening, showing a higher rate of flower and pod abortion. A similar result was found by Calvache et al. (2007), studying bean plants, where stress during the flowering period (R5) and pod formation (R6) are the most sensitive. This shows that a lack of water during the flowering and pod formation periods has a notable influence on the number of pods per plant, given that an adequate level of water in the soil induces optimum flowering, pod formation and grain filling. In turn, water stress during the flowering and pod formation period causes flowers and young pods to abort, as there is a source-drain competition, which results in the elimination of abnormal flowers that fail to fertilise, or younger pods that abort due to a lack of nitrogen or carbohydrates, as also found by Hostalácio and Valio (1984). Water stress during the pod filling stage results in the abortion of young pods and the production of shrivelled pods at the tips, given that pod filling occurs from the base to the tips.

The difference in the number of pods of fava beans found in various studies suggests that there is great variability in this characteristic in relation to the varieties and the soil and climate conditions of the different growing regions. The number of pods per plant is affected by fertilisation and water stress. Fageria et al. (2003) and Navarro Júnior and Costa (2002) state that the number of pods per plant is the most important component when seeking increases in grain yield. Oliveira et al. (2011), when evaluating eight fava bean accessions, found that the most productive accessions had the highest number of pods per plant and the least productive accessions had the lowest number of pods. Acosta-Gallegos and Shibata (1989) found a reduction in all production components when beans were subjected to water stress. The reduction in production was greater (42 to 50 per cent) when the stress was applied in the reproductive phase compared to the vegetative phase, and this was attributed to a decrease in leaf area and the number of pods per plant.

Table 8. Summary of the analysis of variance for 100-grain mass, number of pods per plant and number of grains per pod of fava beans under water stress by development stage and full irrigation.

F. V.	Mass of 100 grains	Number of pods per plant	Number of grains per pod
	QM	QM	QM
Treatments	53.71 ns	45,21**	0.21 ns
Waste	26,66	3,50	0,24

ns not significant (p >= .05)
** significant at 1% probability level (p < .01)

Table 9. Values for 100-grain mass, number of pods per plant and number of grains per pod of fava beans under water stress and full irrigation.

TREATMENTS	Mass of 100 grains	Number of pods per plant	Number of Grains per Pod
E. H. in phase (I)	36,61 a	6.37 bc	1,73 a
E. H. in phase (II)	30,24 a	9.62 ab	1,35 a
E. H. in phase (III)	36,55 a	3,75 c	1,87 a
E. H. in phase (IV)	30,32 a	6.00 bc	1,42 a
E. H. in phases (I and II)	25,60 a	6.12 bc	1,46 a
E. H. in phases (II, III and IV)	31,82 a	4,00 c	1,56 a
E. H. in phases (III and IV)	33,79 a	4,37 c	1,97 a
E. H. in phases (I, II and III)	30,59 a	2,50 c	1,55 a
E. H. in phases (I, II, III and IV)	29,60 a	3,12 c	1,64 a
With full irrigation	26,73 a	13,50 a	1,99 a
	CV= 16.56%	CV= 31.54	CV=29.65%
	MG=31.18	MG=5.93	MG=1.65

* Averages followed by the same letter do not differ according to Tukey's test at a 0.05 probability level.

In relation to the analysis of variance for the number of grains per pod, Table 8, according to the F test at 5% probability, there was no significant effect of the treatments water stress by development stage and full irrigation. The average values of the treatments for the number of grains per pod can be seen in Table 9. The average between the treatments was 1.8 grains per pod. The data found in this research is close to that obtained by Oliveira et al. (2011), whose average number of seeds per pod was 2 seeds in the 8 accessions of fava beans studied. Guimarães et al. (2007) found values for the number of seeds per pod ranging from 2 to 6, in a trial with 14 fava bean accessions.

Table 10 summarises the analysis of variance for fava bean pod yield, grain yield and MST production. The results indicate a significant difference for the variables studied, at the 1% level (p <0.01) by the F test. The treatments that show the highest yields and are least sensitive to water stress are: E.H. in phase (I), E.H. in phase (II), E.H. in phase (IV) and E.H. in phases (I and II). H. in phases (I and II) for pod yield, with yields of 1298.53; 1536.94; 1222.81 and 1151.93 kg.ha^{-1} respectively. The treatments most sensitive to water stress with the lowest yields were: E. H. in phases (I, II and III), E. H. in phases (I, II, III and IV) with yields of 296.07 and 539.85 kg.ha^{-1} of fava bean pods respectively. The highest yield was obtained in the treatment under full irrigation, with 2584.98 kg.ha^{-1} of pods.

For grain yield, the highest value was obtained in the treatment under full irrigation -

1407.94 kg.ha^{-1} of grain. Melo (2005), studying fava beans, obtained yields of over 2500 kg.ha^{-1} for an irrigation margin of 761 mm during the fava bean cycle in Sapé-PB, using the "Boca-de-moça" variety. The treatments under stress that show the highest yields and are least sensitive to water stress are: E.H. in phase (I), E.H. in phase (II), E.H. in phase (IV) and E.H. in phases (I and II). H. in phases (I and II) with yields of 643.24; 712.02; 645.65 and 577.23 kg.ha^{-1} respectively of fava bean grains. The treatments most sensitive to water stress with the lowest yields are: E. H. in phases (I, II and III), E. H. in phases (I, II, III and IV) and E. H. in phases (III, III and IV) with yields of 282.79; 311.97 and 357.19 kg.ha^{-1} respectively of fava bean grains. Sensitivity to water stress for grain yield was greater the longer the duration of the water stress (when it occurred in more than one stage of development) and when it occurred during flowering, pod formation and pod filling. According to Fancelli and Dourado Neto (1999), water deficiency during the pod formation stage affects bean yields by reducing the rate of photosynthesis and the plant's metabolism, and by causing pods to fall and grow less. In addition to the direct effect on crop development, water deficit leads to a reduction in the absorption efficiency of some nutrients such as nitrogen and phosphorus.

In terms of MST production, the treatments with the highest production while being less sensitive to water stress are: E. H. in phase (IV), E. H. in phase (III), E. H. in phase (III and IV) with MST production of 1373.19; 1165.9 and 1027.38 kg.ha^{-1} respectively. The treatments most sensitive to water stress with the lowest yields were: E. H. in phases (II); E. H. in phases (I); E. H. in phases (I, II, III and IV); E. H. in phases (I and II) and E. H. in phases (II, III and IV) with yields of 573.49; 725.25; 795.25; 852.99 and 868.71 kg.ha^{-1} of MST respectively (Table 11). Periods of drought produce an increase in root development and accumulation of photoassimilates in the roots, which increases water absorption from soil layers that have water available. Deficit irrigation may be acceptable in areas that have water deficiency at this time. On the other hand, deficient irrigation throughout the cycle drastically reduced the production of pods, grains and MST compared to the treatment without water stress.

Table 10. Summary of the analysis of variance for pod yield, grain yield and total dry matter production (kg ha^{-1}) of fava beans under water stress by development stage and full irrigation.

	Pod yield	Grain yield	Total dry matter production
	QM	QM	QM
Treatments	1746055,53**	422798,3**	17498143,0**
Waste	19845,410	2940,82	67800,29

** significant at 1% probability level (p < .01)

Table 11. Average values for pod yield, grain yield and total dry matter production in water-stressed and unstressed fava beans by development stage.

TREATMENTS	Productivity (kg/ha)		
	Pods	Grains	Total dry matter
E. H. in phase (I)	1298.53 bc	643.24 bc	725.50 of
E. H. in phase (II)	1536,94 b	712,02 b	573,49 e
E. H. in phase (III)	759,62 d	456.52 of	1165.90 bc
E. H. in phase (IV)	1222.81 bc	654.65 bc	1373,19 b
E. H. in phases (I and II)	1151,93 c	577.23 cd	852.99 cde
E. H. in phases (II, III and IV)	630,65 d	357.19 eff	868.71 cde
E. H. in phases (III and IV)	728,77 d	461.05 of	1027.38 bcd
E. H. in phases (I, II and III)	296,07 e	282,79 f	978.25 cd
E. H. in phases (I, II, III and IV)	539.85 of	311,97 f	795.50 of
With full irrigation	2584,98 a	1407,94 a	5267,07 a
	CV= 13.14%	CV= 9.25%	CV= 10.97%
	MG=1072.01	MG=586.46	MG=1747.31

** Averages followed by the same letter do not differ according to Tukey's test at a 0.01 probability level.
4.   4Response factor of fava beans to water deficit (Ky)

According to Dorenboos and Kassam (1994), crops can be classified according to their sensitivity to water stress in four categories: low (Ky < 0.85); low/medium (0.85 < Ky < 1.00); medium/high (1.00 < Ky < 1.15) and high (Ky > 1.15). Analysing the ky averages for pod yield in the treatments: water stress in phase I, water stress in phase II, water stress in phase III and water stress in phase IV; which were subjected to water deficit in just one phenological stage, it can be seen that the relative reduction in fava bean yield was most pronounced in the flowering and pod formation stage (water stress in phase III) with ky = 1.41.

Water deficit in the treatments water stress in phase I, water stress in phase II and water stress in phase IV; vegetative growth and grain filling stages, whose ky, 0.99; 0.81 and 1.05 respectively, had less effect on crop yield, with the least sensitivity to water stress in phases II and I with ky 0.81 and 0.99 respectively (Table 12). These values are higher than those found by Doorenbos and Kassam (1994) who, studying bean plants, obtained ky = 1.1 at flowering, while ky = 1.41 was obtained in the present study; grain filling (ky = 0.75 for Doorenbos and Kassam, compared to ky = 1.05) and vegetative (ky = 0.2 for Doorenbos and Kassam, compared to ky = 0.8 for the present study with fava beans).

The values of the sensitivity coefficients ky in the treatments with water stress in more than one phenological stage water stress in phase (I and II); water stress in phase (III and IV); water stress in phase (I, II and III); water stress in phase (II, III and IV) and water stress in phase (I, II, III and IV), whose ky, 1.10; 1.4; 1.79; 1.51 and 1.58 respectively, with the highest ky for the treatments water stress in phase (I, II and III) water stress during vegetative development and flowering, and also the treatment in which the fava beans were subjected to water deficiency throughout the crop cycle.

In the treatments with water stress in two phases, the highest ky value occurred in the treatment with water stress during pod formation and filling, with a ky of 1.32. Similar results were found by
(CORDEIRO et al., 1998), studying the effect of water stress on cowpea, where water deficiency in the vegetative and flowering stages was greater with ky = 0.66. In turn, the treatments, E. H. in phase (II); E. H. in phase (IV); E. H. in phase (I) and E. H. in phases (I and II) had the lowest ky 0.98; 1.07; 1.08 and 1.18 respectively, showing greater resistance to water stress in these phases, at the beginning of vegetative development and at the end of the reproductive phase when most of the pods are already formed. Doorembos and Kassam (1979), analysing various experiments with bean plants, noted that a 50% water deficit in the vegetative stage causes a yield reduction of only 10%. The same deficit during flowering reduces production by 55 per cent and during pod filling by 38 per cent. However, a deficit during ripening reduces yield by only 10 per cent.

Analysing the average values of sensitivity to water stress (ky) of the treatments for grain production (Table 13), the highest ky values were found in the treatments; E. H. in phases (I, II and III), E. H. in phases (I, II, III and IV) and E. H. in phases (II, III and IV). H. in phases (II, III and IV) with ky 1.59; 1.55 and 1.49 respectively, showing high sensitivity to water stress. It can be seen that the longer the duration of water stress, the greater the Ky, i.e. when the duration of water stress is prolonged over time during the crop's development phases, the greater its effect on reducing production.

Water stress during flowering and ripening, the treatments, E. H. in phase (III) and E. H. in phases (III and IV) also show high ky 1.35 and 1.34 respectively. This shows that the fava bean crop is very sensitive to water stress during these phenological phases as it causes flowers and pods to abort. Similar results were found by (CORDEIRO et al., 1998), studying the effect of water stress on cowpeas, where water deficiency in the vegetative and flowering stages was greater with ky = 0.66. In turn, the treatments, E. H. in phase (II); E. H. in phase (IV); E. H. in phase (I) and E. H. in phases (I and II) had the lowest ky 0.98; 1.07; 1.08 and 1.18 respectively, showing greater resistance to water stress in these phases, at the beginning of vegetative development and at the end of the reproductive phase when most of the pods are already formed.

Doorenbos and Kassam (1979), analysing various experiments with bean plants, noted that a water deficit of 50% in the vegetative stage causes a yield reduction of only 10%. The same deficit during flowering reduces production by 55 per cent and during pod filling by 38 per cent. However, a deficit during ripening reduces yield by only 10 per cent.

Table 12. Average values of [1-(ETr/ETm)] and [1-(yr/ym)] and the sensitivity factor ky of fava beans, for pod yield (kg.ha^{-1}), in treatments under water deficit.

TREATMENTS	ETr	ETm	ETr/ETm	(1-ETr/ETm)	Yr	Ym	Yr/Ym	(1-Yr/Ym)	Ky
E. H. in phase (I)	10,67	21,35	0,5	0,5	1298,53	2584,98	0,5023	0,4977	0,9953
E. H. in phase (II)	29,76	59,51	0,5	0,5	1536,94	2584,98	0,5946	0,4054	0,8109
E. H. in phase (III)	35,43	70,86	0,5	0,5	759,62	2584,98	0,2939	0,7061	1,4123
E. H. in phase (IV)	21,58	43,17	0,5	0,5	1222,81	2584,98	0,4730	0,5270	1,0539
E. H. in phases (I and II)	40,43	80,86	0,5	0,5	1151,93	2584,98	0,4456	0,5544	1,1088
E. H. in phases (II, III and IV)	86,77	173,54	0,5	0,5	630,65	2584,98	0,2440	0,7560	1,5121
E. H. in phases (III and IV)	57,01	114,03	0,5	0,5	728,77	2584,98	0,2819	0,7181	1,4362
E. H. in phases (I, II and III)	75,86	151,72	0,5	0,5	266,07	2584,98	0,1029	0,8971	1,7941
E. H. in phases (I, II, III and IV)	97,44	194,89	0,5	0,5	539,85	2584,98	0,2088	0,7912	1,5823

* ETr, ETm: Actual and maximum evapotranspiration, in mm.day^{-1} , respectively.

Yr, Ym: Actual and maximum yield, in kg.ha^{-1} of fava bean pods, respectively

Table 13. Average values of [1-(ETr/ETm)] and [1-(yr/ym)] and the sensitivity factor ky of fava beans, for grain yield (kg.ha^{-1}), in treatments under water deficit.

TREATMENTS	ETr	ETm	ETr/ETm	(1-ETr/ETm)	Yr	Ym	Yr/Ym	(1-Yr/Ym)	Ky
E. H. in phase (I)	10,67	21,35	0,5	0,5	643,24	1407,94	0,4569	0,5431	1,0863
E. H. in phase (II)	29,76	59,51	0,5	0,5	712,02	1407,94	0,5057	0,4943	0,9886
E. H. in phase (III)	35,43	70,86	0,5	0,5	456,52	1407,94	0,3242	0,6758	1,3515
E. H. in phase (IV)	21,58	43,17	0,5	0,5	654,65	1407,94	0,4650	0,5350	1,0701
E. H. in phases (I and II)	40,43	80,86	0,5	0,5	577,23	1407,94	0,4100	0,5900	1,1800
E. H. in phases (II, III and IV)	86,77	173,54	0,5	0,5	357,19	1407,94	0,2537	0,7463	1,4926
E. H. in phases (III and IV)	57,01	114,03	0,5	0,5	461,05	1407,94	0,3275	0,6725	1,3451
E. H. in phases (I, II and III)	75,86	151,72	0,5	0,5	282,79	1407,94	0,2009	0,7991	1,5983
E. H. in phases (I, II, III and IV)	97,44	194,89	0,5	0,5	311,97	1407,94	0,2216	0,7784	1,5568

* ETr, ETm: Actual and maximum evapotranspiration, in mm.day$^-$, respectively.

Yr, Ym: Actual and maximum yield, in kg.ha^{-1} of fava bean grains, respectively

The summary of the analyses of variance for water use efficiency is shown in Table 14. There was a significant difference at the 1% probability level by the F test for water use efficiency, pod yield, grain yield and total dry matter.

Table 14. Summary of the analysis of variance for water use efficiency.

FV	Water use efficiency		
	Pods	Grains	Total dry matter.
	GL	QM	QM	QM
Treatments	9	0,270**	0,059**	1,296**
Waste	30	0,020	0,0056	0,0066

** significant at 1% probability level (p < .01)

The mean test for water use efficiency for pod, grain and MST production of fava beans is shown in Table 14. For pod production, the treatment with full irrigation showed the highest WUE of 1.130 kg.m^{-3} ; in the treatments under water stress, those that showed the highest water use efficiency were: E. H. in phase (II), E. H. in phases (I and II), E. H. in phases (I) and E. H. in phase (IV), with 0.783; 0.644; 0.614 and 0.612 kg.m$^-$ 3 respectively. The treatment with the lowest US was E. H. in phases (I, II and III) with 0.180 kg.m3.

In terms of grain production, the treatment with full irrigation showed the highest EUA of 0.615 kg.m$^-$ 3; in the treatments under water stress, the ones that showed the highest water use efficiency were: E. H. in phase (II), E. H. in phases (I and II), E. H. in phase (IV) and E. H. in phase (IV). H. in phase (I), with 0.363; 0.322; 0.328 and 0.304 kg.m$^-$ 3 respectively. The treatment with the lowest EUA was E. H. in phases (I, II and III) with 0.165 kg.m$^-$ 3 . The US values found are in line with those recommended for beans by Doorenbos and Kassan (1979), in the order of 0.30 to 0.60 kg m^{-3} for grains with a 10% moisture content. Similar results were found by Calvache et al. (1997) who studied the water use efficiency of the INIAP 4040 bean cultivar and obtained a water use efficiency ranging from 0.46 to 0.92 kg m^{-3} in treatments with different irrigation rates and nitrogen dosages; and by Barros and Hanks (1993) who found maximum values of 0.65 kg m^{-3} and 0.75 kg m^{-3} for the water use efficiency based on the dry matter of bean grains.

For MST production, the treatment with full irrigation showed the highest US 2.303 kg m^{-3} ; in the treatments under water stress, the ones that showed the highest water use efficiency were: E. H. in phases (I, II, III and IV), E. H. in phase (IV), E. H. in phases (I, II and III), E. H. in phases (II, III and IV), E. H. in phases (III and IV), E. H. in phases (III and IV) with 0.695; 0.688; 0.681; 0.660; 0.625 and 0.603 kg.m$^-$ 3 respectively. The treatments with the lowest EUA were: E. H. in phase (II), E. H. in phase (I) with 0.292; 0.343 kg.m$^-$ 3 respectively. These values differ from those found by Barros and Hanks (1993) for water use efficiency, based on biomass, with 1.17 and 1.41 kg m^{-3} for the treatments with bare soil and mulch respectively.

Table 15. Mean test for water use efficiency for pod, grain and MST production of fava beans (kg ha$^-$

¹), in treatments under water deficit and full irrigation.

Treatments	Water use efficiency (kg.m3)		
	Pods	Grains	Total dry matter
E. H. in phase (IV)	0.525 bcd	0.296 bc	0,688 b
E. H. in phase (III)	0.392 cd	0.249 bc	0.603 bc
E. H. in phases (I and II)	0.580 bc	0.319 bc	0.477 cd
E. H. in phases (II, III and IV)	0.698 bc	0,414 b	0.660 bc
E. H. in phases (I, II and III)	0,180 d	0,165 c	0,681 b
E. H. in phases (I, II, III and IV)	0.428 cd	0.263 bc	0,695 b
E. H. in phase (II)	0.847 ab	0,404 b	0,292 d
E. H. in phase (I)	0.595 bc	0,365 b	0,343 d
E. H. in phases (III and IV)	0.649 bc	0.340 bc	0.625 bc
With full irrigation	1,130 a	0,615 a	2,303 a
	Mg= 0.602	Mg=0.343	Mg=0.737
	CV%=27.31	CV%=21.87	CV=11.05

** Averages followed by the same letter do not differ according to Tukey's test at a 0.01 probability level.

CHAPTER 5

CONCLUSIONS

Considering the conditions in which the work was carried out, the results obtained allow us to conclude that:

- Water deficit decreased the leaf area index, chlorophyll content and number of pods per plant and increased flower and pod abortion.

- The fava bean was more sensitive to water stress when it occurred at more than one stage of development (over a longer period of time) and when it occurred during flowering and pod formation; it was more tolerant during vegetative development and pod ripening.

- Water deficit affected fava bean yields in terms of pods, grains and total dry matter.

- The efficiency of water use by the fava bean was higher in the treatments without water stress.

CHAPTER 6

BIBLIOGRAPHICAL REFERENCES

ABOUKHALED, A.; ALFARO, A.; SMITH, M. Lysimeters. Rome: FAO, Irrigation and Drainage Paper, 39, 1982. 68p.

ACOSTA-GALLEGOS, J.A.; SHIBATA, J.K. Effects of water stress on growth and yield of indeterminate dry bean (Phaseolus vulgaris L.) cultivars. Field Crop Research, v.20, p.81-93, 1989.

ALLEN, R. G. et al. Crop evapotranspiration: guidelines for computing crop water requirements. Rome: FAO, 2006. 326 p. Irrigation and Drainage, 56.

ALLEN,R.G.; PEREIRA,L.S.; RAES,D.; SMITH,M. Crop evapotranspiration - guidelines for computing crop water requirements. Rome: FAO, 1998, 300p. (FAO, Irrigation and drainage paper 56).

ALVARENGA, I. C. A. Water stress in pepper rosemary (Lippia Sidoides Cham.): physiological and productive aspects - 2010. 61 f. Master's thesis - Federal University of Minas Gerais - UFMG, Montes Claros, MG: ICA/UFMG, 2010.

ANDRADE JÚNIOR, A.S.; BASTOS, E.A.; BARROS, A.H.C.; SILVA, C.O.; GOMES, A.A.N. Climatic classification and regionalisation of the semi-arid region of the State of Piauí under different rainfall scenarios. Revista Ciência Agronômica, v. 36 , n. 2, p. 143 - 151, 2005.

ANGELOCCI, L.R. Water in the plant and gas/energy exchange with the atmosphere: introduction to biophysical treatment. Piracicaba: The author, 2002. 272 p.

ANGIOI, S.A.; DESIDERIO, F.; RAU, D.; BITOCCHI, E.; ATTENE, G.; Papa, R. Development and use of chloroplast microsatellites in Phaseolus spp. and other legumes. Plant Biology, v.11, p.598-612, 2009.

ANTONY, E.; SINGANDHUPE, R.B. Impact of drip and surface irrigation on growth, yield and WUE of capsicum (Capsicum annuum L.). Agricultural Water Management 65, p. 121-132. 2004.

ARAÚJO, R. C. P. Avaliação de alternativas tecnológicas para a cajucultura do Nordeste sob condições de risco. Fortaleza, UFC, 1992 (Master's dissertation).

AZEVEDO FILHO, A. J. B. V. Economic analysis of projects: Software for deterministic and risk situations involving simulation. Piracicaba: ESALQ/USP, 1988 (Master's dissertation).

AZEVEDO, J. N.; FRANCO, L. J. D.; ARAÚJO, R. O.C.; Chemical composition of seven varieties of fava beans. Piauí: EMBRAPA/CNPMN, 2003. 4p. (Technical communication, 152)

BERLATO, M. A.; MOLION, L. C. B. Evaporation and evapotranspiration. Porto Alegre: IPAGRO, 1981. 95p. (IPAGRO. Technical Bulletin, 7).

BERNARDO, S.; SOARES, A. A.; MANTOVANI, E. C. Manual de irrigação. 7.ed. Viçosa: Editora UFV, 2005. 611 p.

BERTONHA, A. Response functions of pear oranges to supplementary irrigation and nitrogen. Piracicaba, 1997. 113p. Thesis (Doctorate) - Luiz de Queiroz College of Agriculture, University of São Paulo.

BEYRA, A.; ARTILES, G. R. Taxonomic revision of the genera Phaseolus and Vigna (Leguminosae - Papilionoideae) in Cuba. Anales Del Jardín Botánico de Madrid. v.61, n.2, p.135-154, 2004.

BEZERRA, F.M.L.; MESQUITA, T.B. de. Maximum evapotranspiration and crop coefficients of chilli grown in drainage lysimeters. Horticultura Brasileira, Brasília, v.18, suplemento/junho, p. 600-601, 2000.

BLANEY, H. F.; CRIDDLE, W. D. Determining water requirements in irrigated areas from climatological and irrigation data. Washington: USDA, 1950. 48p.

BRADY, NYLE C. Nature and properties of soils. 7. ed., Freitas Bastos: Rio de Janeiro, 1989. p.56-62.

BRAGAGNOLO,N.; MIELNICZUK,J. Soil cover with wheat straw and its relationship with soil temperature and humidity. Revista Brasileira de Ciência do Solo, Viçosa,v.14,n.3p.369-374,1990.

BRIGGS, G.E.; KID, F.; WEST, C. A quantitative analysis of plant growth. Part l. Annals of Applied Biology, v.7, p.103-123, 1920.

BRONDANI, R.P.V., BRONDANI, C., TARCHINI, R., GRATTAPAGLIA, D. Development characterisation and mapping of microsatellite markers in Eucalyptus grandis and E. urophylla. Theoretical and Applied Genetics, v.97, p.816-827, 1998.

BROUGHTON, W.J.; HERNÁNDEZ, G.; BLAIR, M.; BEEPBE, S.; GEPTS, P.; ANDERLEYDEN,

J. Beans (Phaseolus spp.) - model food legumes. Plant and soil, v.252, n.1, p.55-128, 2003.

BROWN, LESTER. A desert full of people. In: The dispute over blue gold, Lmonde Diplomatique notebooks. Anita Garibaldi Publishing House, 2003

BRUTSAERT, W. Evaporation into the atmosphere: theory, history and applications. Dordrecht: Kluver Academic, 1982. 299p. (Environmental Fluid Mechanics, 1).

CONFALONE, A.E.; COSTA, L.C.; PEREIRA, C.R. Growth and light capture in soya beans under water stress. Revista Brasileira de Agrometeorologia, Santa Maria, v.6, n.2, p.165-169, 1998.

CAMPBELL, G.S.; NORMAN, J.M. An introduction to environmental biophysics. New York: Spring-Verlag, 1998. 286 p.

CAMPOS, R. T. Efeitos do ataque do bicudo na cotonicultura do semi-árido cearense. Recife: Federal University of Pernambuco, 1991 (Doctoral Thesis).

CARDOSO, M.J.; MELO, F.B.; ANDRADE JÚNIOR, A.S. Influence of caupi (Vigna Unguiculata (L.) Walp.) plant density on grain yield and its components under irrigation. In: REUNIÃO NACIONAL DE PESQUISA DE CAUPI, 4, 1996, Teresina. Abstracts , Teresina: EMBRAPA/CPMM, 1996. 123 p.

CORDEIRO, L.G.; BEZERRA, F.M.L.; SANTOS, J.J.A ; MIRANDA, E.P. Sensitivity factor to water deficit (ky) of the cowpea crop (Vigna unguiculata (L.) Walp.). In: BRAZILIAN CONGRESS OF AGRICULTURAL ENGINEERING, 27, 1998. Poços de Caldas, MG. Proceedings , volume II, Poços de Caldas, MG: Sociedade
Brasileira de Engenharia Agrícola, 1998.p. 178-180.

COUTO, L; SANS, L.M.A. Requerimento de água das culturas. Sete Lagoas: Embrapa, 2002. 10 p. (Technical Circular, 20).

CRONQUIST, A. Devolution and classification of flowering plants. New York: New York Botanical Garden, 1988. 555 p.

CUNHA, A. R. da. Agrometeorological parameters of pepper crops (Capsicum annuum L.) in protected and field environments. Thesis (Doctorate) UNESP- Botucatu, 2001. 128 P.

DEBOUCK, D.G. Diversity in Phaseolus species in relation to the common bean. In: SINGH, S. P. (Ed.). Common bean improvement in the twenty-first century. Dordrecht: Kluwer, 1999. p.25-52.

DOORENBOS, J.; KASSAM, A. H. Effect of water on crop yields. Trad. By H.R. Gheyi, A.A. de Sousa, F.A.V. Damasceno and J.F. de Medeiros. Campina Grande: UFPB, 1994. 306p. (FAO. Estudos de Irrigação e Drenagem, 33).

DOORENBOS, J.; PRUITT, J.O. Guidlines for predicting crop water requirements. Rome, FAO, 1977. 179p. (FAO Irrigation and Drainage Paper, 24)

DOURADO NETO, D.; LIER, Q.J.V.; BOTREL, T.A.; LIBARDI, P.L. Soil water retention curve: quickbasic algorithm for estimating the empirical parameters of the Genuchten model. Piracicaba: ESALQ, 1990. 32p. (User's manual).

DURIGON, A. Soil-plant-atmosphere water transfer mechanisms and their relation to crop water stress. Thesis (Doctorate). ESALQ, Piracicaba, 2011. 143 p.

ENGEL, V. L.; POGGIANI, F. Study of the chlorophyll concentration in leaves and their light absorption spectrum as a function of shading in seedlings of four

native forest species. Revista Brasileira de Fisiologia Vegetal, v.3, p.3945, 1991.

ERTEK, A.; SENSOY, S.; GEDIK, I. Irrigation scheduling based on pan evaporation values for cucumber (Cucumis sativus L.) grown under field conditions. Elsevier Science Agricultural Water Management. 14p.2005

FAGERIA, N. K.; BARBOSA FILHO, M. P.; STONE, L. F. Bean plant response to phosphate fertilisation. In: POTAFÓS. Symposium highlights the essentiality of phosphorus in Brazilian agriculture. Informações Agronômicas, Piracicaba - SP, n.102, p.1-9, 2003.

FALKER. CFL1030 - chlorofiLOG - Electronic chlorophyll meter. Available at: http://www.falker.com.br/. Accessed on;18 August. 2012.

FARIAS, J.R.B., BERGAMASCHI, H., MARTINS, S.R. Evapotranspiration inside plastic greenhouses. Revista Brasileira de Agrometeorologia, Santa Maria, v.2, p.17-22, 1994.

FILGUEIRA, F.A.R. Novo manual de olericultura: Agrotecnologia moderna na produção e comercialização de hortaliças. Viçosa: UFV, p.235-9, 2000.

FOLEGATTI, M.V.; PAZ, V.P.S.; PEREIRA, A.S.; LIBARDI, V.C.M. Effect of different levels of irrigation and water deficit on bean production (Plaseolus vulgaris L). In: CONGRESSO CHILENO DE ENGENIERIA AGRÍCOLA, 2., 1997, Chillán. Floppy disc. Chillán, 1997.

FOLEGATTI, M.V.; SCATOLINI, M.E.; PAZ, V.P.S. et al. Effects of the plastic cover on the meteorological elements and evapotranspiration of the chrysanthemum crop in a greenhouse. Revista Brasileira de Agrometeorologia, Santa Maria, v.5, n.2, p.155-163, 1997.

FRANCO, M.C.; CASSINI, S.T.A.; OLIVEIRA, V.R.; VIEIRA, C. & TSAI, S.M. Nodulation in bean cultivars from the Andean and Mesoamerican gene pools. Pesquisa Agropecuária Brasileira, 37:1145-1150, 2002.

FREIRE, A.L.D.O. Effects of water deficit on some biophysical and biochemical aspects and on the development of runner beans (Phaseolus vulgaris L.). Lavras: ESAL, 1990. 86 p. (Master's thesis in Plant Physiology).

FREITAS, J. A. G. Hydraulic load sensor evapotranspirometer: construction, calibration and tests. Recife, 1994. 122p. Dissertation (Master's Degree) - Federal Rural University of Pernambuco.

FREYTAG, G.F.; DEBOUCK, D.G. Taxonomy, Distribution, and Ecology of the Genus Phaseolus (Leguminosae-papilionoideae) in North America, Mexico and Central America. Botanical Research Institute of Texas (BRIT), Forth Worth, TX, USA. 2002. 298p.

GAITÁN-SOLÍS, E.; DUQUE, M.C.; EDWARDS, K.J.; TOHME, J. Microsatellite repeats in common bean (Phaseolus vulgaris L.): isolation, characterisation, and cross-species amplification in Phaseolus ssp. Crop Science v.42, n.6, p.2128-2136, 2002.

GAVANDE, S.A. Física de suelos, princípios e aplicaciones. 2 ed. Mexico: Ed. Limusa, 1976. 351 p.

GONZALES, A.R.; WILLIAMS, J.W. Effect of water stress during pod development on yield and quality of raw and canned sanp beans. HortScience, Virginia, v. 14, n. 2, p. 125, Apr. 1979.

GUIMARÃES, W. N. R. et al. Morphological and molecular characterisation of fava bean accessions (Phaseolus lunatus L.). Revista Brasileira de Engenharia Agrícola e Ambiental, v. 11, n. 01, p. 37-45, 2007.

HARDY O.; DUBOIS, S.; ZORO BI, I.; BAUDOIN, J.P. Gene dispersal and its consequences on the genetic structure of wild populations of Lima bean (Phaseolus lunatus) in Costa Rica. Plant Genetic Resources Newsletter, n.109, p.1-6, 1997.

HANKS, R.J.; SHAWCROFT. An economical hydraulic weighing evaporation tank. Transaction of the ASAE, v.16, n.2, p.294-295, 1965.

HENDRY, G. A. F.; PRICE, A. H. Stress indicators: chlorophylls and carotenoids.
In:Hendry, G. A. F.; Grime, J. P. (eds), Methods in Comparative Plant Ecology, p. 148152. London,
Chapman & Hall, 1993.

HOSTALÁCIO, S. Study of some physical, biochemical and anatomical aspects in the growth and
development of beans under different irrigation regimes. Campinas: UNICAMP, 1984. 144 p.
(Thesis. Doctorate in Plant Biology)

HOSTALACIO, S.; VALIO, I.F.M. Development of bean fruits under different irrigation regimes.
Pesquisa Agropecuária Brasileira, v.19, n.1, p.53-57, 1984.

HULUGALLE, N.R.; WILLATT, S.T. The role of soil resistance in determining water uptake by
plant root systems. Australian Journal of Soil Research, Melbourne, v. 24, n. 4, p. 571-574, Sept.
1983.

BRAZILIAN INSTITUTE OF GEOGRAPHY AND STATISTICS - IBGE. Municipal Livestock
Survey. 2010. Available at: < http://www.sidra.ibge.gov.br/bda/ tabela/listabl.asp?c=73;z=p;o=27>.
Accessed on: 30 June 2011.

JARVIS, A.J.; DAVIES, W.J. The coupled response of stomatal conductance to photosynthesis and
transpiration. Journal of Experimental Botany, Lancaster, v. 49, p. 399-406, June 1998.

JONG VAN LIER, Q. DE; VAN DAM, J.C.; METSELAAR, K.; JONG, R. DE; DUIJNISVELD,
W.H.M. Macroscopic root water uptake distribution using a matric flux potential approach. Vadose
Zone Journal, Madison, v. 7, p. 1065-1078, Aug. 2008.

KARAMANOS, A.J.; ELSTON, J.; WADSWORTH, R.M. Water stress and leaf growth of field
beans (Vicia faba L.) in the field: water potentials and laminar expansion. Annals of Botany, New
York, v. 49, n. 6, p. 815-826, June 1982.

KRAMER, P. J., Boyer, J. S. Water relations of plants and soils. San Diego: Academic Press, 1995.
495p

KUIPER, P.J.C. The effects of environmental factors on the transpiration of leaves, with special
reference to stomatal light response. Landbouwhoge School Wageningen, Wageningen, v. 7, p. 1-49,
Nov. 1961.

LABANAUSKAS, C. K.; SHOUSE, P.; STOLZY, L.H. Effects of water stress at various growth stages on seed yield na nutrient concentrations of fieldgrowncowpeas. Soil Science, Baltimore, v. 131, n. 4, p. 249-256, 1981.

LEE, J.M.; READ, P.E.; BADIS, D.W. Effect of irrigation on interlocular cavitation on yield in snap bean. Journal of the American Society for Horticultural Science, Mount, v. 102, n. 3, p. 276- 278, May 1977.

LETEY, J. Relationship between soil physical properties and crop production. In: Advances in soil science.California: Springer-Verlag New York, v.1.p.277- 293,1985.

LIMA, C. J. G. S. et al. Mathematical models for estimating caupí bean leaf area. Revista Caatinga, Mossoró, v. 21, n. 1, p. 120-127, 2008.

MAGALHÃES, A.A.; MILLAR, A.A.; CHOUDHURY, E.N. Effect of phenological water deficit on bean production. Turrialba, Turrialba, v. 29, n. 4, p. 269-273, Oct/Dec./979.

MAGALHÃES, A. C. N. Fotossíntese. in: Fisiologia Vegetal. FERRI, M. G. (ed.) Editora Pedagógica Universitária. São Paulo. p.117180,

MAQUET, A.; VEKEMANS, X.Z.; BAUDOIN, J.P. Phylogenetic study on wild allies of lima bean, Phaseolus lunatus L. (Fabaceae), and implications on its origin. Plant Systematics and Evolution, v.218, n.1-2, p.43-54, 1999.

MAQUET, A.; ZORO BI, I.; DELVAUX, M.; WATHELET, B.; BAUDOIN, J. P. Genetic structure of a Lima bean base collection using allozyme markers. Theoretical and Applied Genetics, v.95, p.980-991, 1997.

MARTINS, G. Use of the greenhouse with plastic cover in *summer* tomato growing. Jaboticabal, 1992. 65p. Thesis (Doctorate) - Faculty of Agricultural and Veterinary Sciences - São Paulo State University.

MARSHALL, J. K. Methods of leaf area measurement of large and small leaf samples. Photosynthetica, v.2, p.4147,1966.

McFARLAND, M.J.; WORTHINGTON, J. W.; NEWMAN, J. S. Design, installation and operation of a twin weighing lysimetre for fruit trees. Transactions of the ASAE, v.26, n.6, p.1717-1721, 1983.

MELCHIOR, H. A Engler's syllabus der pflanzenfamilien. 12. ed. Berlin: Gebruder Borntrager, 1964. 666p.

MELO, L. J. V. Morphophysiology and yield of fava beans under different cultural management conditions. Thesis (Thematic Doctorate in Natural Resources). Federal University of Campina Grande, Campina Grande, 2005. 166p

MONTEIRO, J. E. B. A. et al. Estimation of cotton leaf area using leaf dimensions and mass. Bragantia, Campinas, v. 64, n. 1, p. 15-24, 2005.

MONTEITH, J.L. A reinterpretation of stomatal responses to humidity. Plant, Cell and Environment, Logan, v. 18, n.4, p. 357-364, June 1995.

MONTERO, J.I., CASTILLA, N., GUTIERREZ de RAVÉ, E., BRETONES, F. Climate under plastic in the Almeria area. Acta Horticulturae, Wageningen, v.170, p.227- 234, 1985.

MOUSINHO, F.E.P. Irrigation. In: LOPES, A. C. A.; GOMES, R. L. F.; ARAUJO, A. S. F. The cultivation of fava beans in the mid-north of Brazil. Teresina : EDUFPI, 2010. p.157-171.

NAVARRO JÚNIOR, H. M.; COSTA, A. C. Relative contribution of growth components to soya bean production. Pesquisa Agropecuária Brasileira, Brasília, v. 37, n. 2, 2002.

NEGRI, V.; TOSTI, N. Phaseolus genetic diversity maintained on-farm in central Italy. Genetic Resources and Crop Evolution, v.49, p.511-520, 2002.

NORONHA, J. F. Projetos agropecuários: administração financeira, orçamentária e viabilidade econômica. 2. ed. São Paulo: Atlas, 1987.

OLIVEIRA, A. D. de.; FERNANDES, E. J.; RODRIGUES, T. de J. D. Stomatal conductance as an indicator of water stress in beans. Engenharia Agrícola, Jaboticabal, v.25, n.1, 2005. p.86-95.

OLIVEIRA, F. N.; TORRES, S. B.; BENEDITO, C. P. Agronomic botanical characterisation of fava bean accessions in Mossoró, RN. Revista Caatinga, v. 24, 2011.

OLIVEIRA, M. C. P. et al. Phenology and vegetative development. In: LOPES, A. C. A.; GOMES, R. L. F.; ARAUJO, A. S. F. The cultivation of fava beans in the mid-north of Brazil. Teresina : EDUFPI, 2010. p.103-115.

OLIVEIRA, M. S. de. Effect of water deficit applied at different stages of the phenological cycle of the bean plant (Phaseolus vulgaris L.). cv. Eriparsa. Lavras: ESAL, 1987. 60 p. (Master's Degree in Plant Science)

OLIVEIRA, J. P. Non-destructive method for determining the leaf area of caupi bean, vigna sinensis (L) savi, grown in a greenhouse. Ciência Agronômica, v.7, n.12,p.5357,1977.

PEREIRA, G.M.; CARVALHO, J.A.; RODRIGUES, L.S.; DOBASHI, A.M. Effects of different levels of water deficit applied at three stages of the phenological cycle of bean (Phaseolus vulgaris, L.) c.v. carioca-MG. In: BRAZILIAN CONGRESS OF AGRICULTURAL ENGINEERING, 27, 1998, Poços de Caldas, MG, 1998.

PHILIP, J.R., Evaporation and moisture and heat fields in the soil. Journal of meteorology. 14: 354-366. 1957

PRADOS, N.C. Contribución al estudio de los cultivos enarenados en Almeria: necessidades hídricas y extración del nutrientes del cultivo de tomate decrescimento

indeterminate in a polythene shelter. Almeria, Spain, 1986. 195p. Thesis (Doctorate in Plant Science) - Caja Rural Provincial.

RAATS, P. Uptake of water from soils by plant roots. Transport in Porous Media, Berlin, v. 68, n. 1, p. 5-28, Aug. 2007.

RAMALHO, M. A. P.; SANTOS, J. B. dos; ZIMMERMANN, M. J. de O. Quantitative genetics in autogamous plants: applications to bean improvement. Goiânia: UFG, 1993. 271 p.

REICHARDT, K. Transfer processes in the soil-plant-atmosphere system. Campinas. Cargill Foundation. 1985. 486p.

REICHARDT, K.; TIMM L. C.; Soil, Plant and Atmosphere: Concepts, Processes and Applications. Barueri : Manole, p.323-340, 2004.

REICHARDT, KLAUS. Water in agricultural systems. São Paulo: Manole, 1990. p.27-37.

RODRIGUES, J. J. V. Construction and preliminary tests of a hydraulic load cell evapotranspirometer: In: CONGRESSO BRASILEIRO DE CIÊNCIA DO SOLO, 21, Campinas, 1987. Proceedings. Campinas: Brazilian Society of Soil Science, 1987. p.68.

ROSENBERG, N. J., McKENNY, M.S., MARTIN, P. Evapotranspiration in greenhousewarmed world: a review and a simulation. Agriculture Forest Meteorology, Amsterdam, v.47, p.303-320, 1989.

SANTOS, D; CORLETT, F. M. F.; MENDES, J. E. M. F.; WANDERLEY JÚNIOR, J. S. A. Productivity and morphology of pods and seeds of fava bean varieties in the state of Paraíba. Pesquisa Agropecuária Brasileira, Brasília, DF, v.37, n.10, p.1407-1412, 2002.

SANTOS, R. F.; CARLESSO, R. Water deficit and the morphological and physiological processes of plants. Revista Brasileira de Engenharia Agrícola e Ambiental, v.2, n.3, p.287294,1998.

SAUNDERS, L. C. U. Methods of determination and spatial variability of hydraulic conductivity under field conditions. Piracicaba: ESALQ. 1978. 71p. PhD Thesis.

SCHRODER, T.; JAVAUX, M.; VANDERBORGHT, J.; KORFGEN, B.; VEREECKEN, H. Effect of local soil hydraulic conductivity drop using a three-dimensional root water uptake model. Vadose Zone Journal, Madison, v. 7, n. 3, p. 1089-1098, Aug. 2008.

SEDIYAMA, G.C. Estimation of evapotranspiration: history, evolution and critical analysis. Revista Brasileira de Agrometeorologia, Santa Maria, RS, v.4, n.1, p.i-xii, 1996.

SEDIYAMA, G.C.; RIBEIRO, A.; LEAL, B.G. Climate-water-plant relations. In: BRAZILIAN CONGRESS OF AGRICULTURAL ENGINEERING, 27, 1998. Wells of Tails. Irrigation Management Symposium. Poços de Caldas: Brazilian Society of Agricultural Engineering, 1998. p.46-85.

SENTELHAS, P.C. Agrometeorology applied to irrigation. In: MIRANDA, J.H., PIRES, R.C.M. Irrigação. Piracicaba: FUNEP, v.1, p.63-120,2001.

SENTELHAS, P.C. Class A pan coefficients (Kp) to estimate daily reference evapotranspiration (ETo). Revista Brasileira de Engenharia Agrícola e Ambiental, Campina Grande, PB, v.7, n.1, p.111-115, 2003.

SEVERINO, L. S.; CARDOSO, G. D.; VALE, L. S.; SANTOS, J. W. Method for determining leaf area in papaya. Revista Brasileira de Oleaginosas e Fibrosas, v.8, n.1, p.753-762, 2004.

SILVA NETO, J. R. Morpho-agronomic characterisation and evaluation of resistance in fava bean accessions to golden mosaic and anthracnose. - 2010. 93 f. Dissertation (Master's in Agronomy) - Federal University of Alagoas, Rio Largo -AL.

SILVA, H.T.; COSTA, A. O. Botanical characterisation of wild species of the genus Phaseolus L. (Leguminosae). Santo Antônio de Goiás, 2003. 40p. (Embrapa Arroz e feijão: Comunicado Técnico, 156).

SILVA, F. de A. S.; AZEVEDO, C. A. V. de. Principal Components Analysis in the Software Assistat-Statistical Attendance. In: WORLD CONGRESS ON COMPUTERS IN AGRICULTURE, 7, Reno-NV-USA: American Society of Agricultural and Biological Engineers, 2009.

SILVA, F.C.; FOLEGATTI, M.V.; MAGGIOTTO, S.R. Analysing the operation of a load cell weighing lysimeter. Revista Brasileira de Agrometeorologia, Santa Maria, v.7, n.1, p.53-58, 1999.

SILVA, L. C. et al. A simple method for estimating the leaf area of sesame plants (Sesamum indicum L.). Revista Brasileira de Oleaginonas Fibrosas, Campina Grande, v. 6, n. 1, p. 491-496, 2002.

SOUZA, R.A.; HUNGRIA, M.; FRANCHINI, J.C.; MACIEL, C.D.; CAMPO, R.J.; ZAIA, D.A.M. Minimum set of parameters for evaluating soil microbiota and biological nitrogen fixation by soya. Pesquisa Agropecuária Brasileira, v.43, n.1, p. 83-91, 2008.

STANGHELLINI, C. Evapotranspiration in greenhouse with special reference to mediterranean conditions. Acta Horticulturae, Wageningen, v.335, p.296-304, 1993.

STEWART, J. I.; HAGAN, R.M. Functions to predict effects of crop water deficits. Journal of the Irrigation and Drainage Division. New York, v. 99, IR4, p. 421-39, 1973.

TANNER, C. B. Measurements of evapotranspiration. In: Irrigation of Agricultural Lands. American Society of Agronomy, n.11, p.534-574, 1967.

TARDIEU, F.; DAVIES, W.J. Stomatal response to abscisic acid is a function of current plant water status. Plant Physiology, Waterbury, v. 48, p. 540-545, Oct. 1992.

TARDIEU, F.; SIMMONNEAU, T. Variability among species of stomatal control under fluctuating soil water status and evaporative demand: modelling isohydric and anisohydric behaviours. Journal of Experimental Botany, Lancaster, v. 49, p. 419 432, Jun. 1998.

THORNTWAITE, G.W. An approach towards a rational classification of climate. Geographycal Rev., New York, v.38, n.1, p.55-94. 1948.

TIVELLI, S.W. Chilli cultivation. In: GOTTO, R.; TIVELLI, S.W. Vegetable production. São Paulo: UNESP,p.225-256.1998.

TORRES NETTO, A; CAMPOSTRINE, E.; OLIVEIRA, J. G.; BRESSAN-SMITH, R. E. Photosynthetic pigments, nitrogen, chlorophyll a fluorescence and SPAD-502 readings in coffee leaves. Scientia Horticulturae, v.104, p.199-209, 2005.

TRINTINALHA, M.A. Evaluation of the time domain reflectometry (TDR) technique in the determination of moisture in Eutrophic Red Nitossolo. State University of Maringá, 2000, 52 p. (Master's dissertation).

TUZET, A.; PERRIER, A.; LEUNING, R. A coupled model of stomatal conductance, photosynthesis and transpiration. Plant, Cell and Environment, Logan, v. 26, p. 1097-1116, May 2003.

TURNER, N.C. Drought resistance and adaptation to water deficits in crop plants. In: Mussel, H. and Staples, R. C. Stress physiology in crops plants. New York, p.34372, 1979.

VAADIA, Y.; RANEY, F.C.; HAGAN, R.M. Plant water deficits and physiological processes. Annual Review of Plant Physiology, Palo Alto, v. 12, p. 265-292, 1961.

VAN DER POST, C.J., VAN-SHIE, J.J., GRAAF, R. Basic problems of water relationship: energy balance and water suply in glasshouses the West-Nertherlands. Acta Horticulturae, Wageningen, v.35, p.13-21, 1974.

VAN DEN BERG, A. K.; PERKINS, T. D. Evaluation of a portable chlorophyll meter to estimate chlorophyll and nitrogen contents in sugar maple (Acer saccharum Marsh.) leaves.Forest Ecology and Management, v. 200, p. 113-117, 2004.

VERNON, A.J.; ALLISON, J.C.S. A method of calculating net assimilation rate. Nature, v.200, p.814, 1963.

VIEIRA, R. F. The cultivation of fava beans. Informe Agropecuário, Belo Horizonte, v.16, n.174, p.30 -37, 1992.

VILLA NOVA, N. A.; REICHARDT, K. Evaporation and evapotranspiration. In: RAMOS, F.; OCCHIPINTI, A.G.; VILLA NOVA, N.A.; REICHARDT, K.; MAGALHÃES, P.C.; CLEARY, R.W. Hydrological Engineering, Rio de Janeiro, ABRH, 1989. Chap. 3, p. 14597 (ABRH Water Resources Collection, 2).

VÁSQUEZ M. A. N. ; FOLEGATTI M. V. ; DIAS N.S.; SOUSA V.F. Effect of the protected environment cultivated with melon on meteorological elements and their relationship with external conditions. Engenharia Agrícola, Jaboticabal, v.25, n.1, p.137-143. 2005.

VILLA NOVA, N.A. & SENTELHAS, P.C. Evapopluviometer: a new system for measuring Class A tank evaporation. In: Proceedings of the XI Agrometeorology Congress - II Latin American Agrometeorology Meeting. Florianópolis: Brazilian Society of Agrometeorology,1999.v.1.p.2496-2502.

VILLA NOVA, N.A. A. et al. Estimation of accumulated degree days above any base temperature, as a function of maximum and minimum temperatures. São Paulo: Institute of Geography, USP, 1972. 8p. (Caderno de ciência da terra, n.30).

WATSON, D.J. Comparative physiological studies on the growth of field crops 1. Variation in net assimilation rate and leaf area. Annals of Botany, v.11, p.41-76, 1947

WATSON, D.J. The dependence of net assimilation rate on leaf area index. Annals of Botany, v.22, p.37-54, 1958

WATSON, D.J. The physiological basis of variation in yield. Advances in Agronomy, v.4, p.101-145, 1952.

WETZEL, M.M.V. DA S.; SILVA, D.B. DA; SILVA, H.T. DA; NETO, L.G.V.P.; FONSECA, J.R. Acervo de recursos genéticos de Phaseolus spp. Conservados à longo prazo. Brasília, 2006. 10p. (Embrapa Recursos Genéticos e Biotecnologia: Boletim de pesquisa e desenvolvimento, 129).

XAVIER, G.R.; MARTINS, L.M.V.; RIBEIRO, J.R.A. & RUMJANEK, N.G. Symbiotic specificity between rhizobia and cowpea accessions of different nationalities. Caatinga, 19: 25-33, 2006.

ZIMMERMANN, M.J.O; TEIXEIRA, M.G. Origin and evolution. In: ARAÚJO, R.S.; RAVA, C.A.; STONE, L.F.; ZIMMERMANN, M.J.O. eds. Common bean cultivation in Brazil. Piracicaba: Brazilian Association for Potash and Phosphate Research (POTAFOS), 786p, 1996.

ZIMMERMANN, U.; MEINZER, F.; BENTRUP, F.W. How does water ascend in tall trees and other vascular plants? Annals of Botany, Oxford, v. 76, p. 545-551, July 1995.

yes
I want morebooks!

Buy your books fast and straightforward online - at one of world's fastest growing online book stores! Environmentally sound due to Print-on-Demand technologies.

Buy your books online at
www.morebooks.shop

Kaufen Sie Ihre Bücher schnell und unkompliziert online – auf einer der am schnellsten wachsenden Buchhandelsplattformen weltweit! Dank Print-On-Demand umwelt- und ressourcenschonend produzi ert.

Bücher schneller online kaufen
www.morebooks.shop

Printed by Books on Demand GmbH, Norderstedt / Germany